TRANSPORTATION DISASTER RESPONSE HANDBOOK

TRANSPORTATION DISASTER RESPONSE HANDBOOK

Jay Levinson
Israel Police (ret.), Jerusalem, Israel

Hayim Granot
Bar Ilan University, Ramat Gan, Israel

ACADEMIC PRESS
An Elsevier Science Imprint

San Diego San Francisco New York Boston
London Sydney Tokyo

This book is printed on acid-free paper.

ACADEMIC PRESS
An Elsevier Science Imprint
Harcourt Place, 32 Jamestown Road, London NW1 7BY, UK
http://www.academicpress.com

ACADEMIC PRESS
An Elsevier Science Imprint
525 B Street, Suite 1900, San Diego, California 92101-4495, USA
http://www.academicpress.com

ISBN 0-12-445486-0

Library of Congress Catalog Number: 2001093302
A catalogue record for this book is available from the British Library

Typeset by Kenneth Burnley, Wirral, Cheshire
Printed by Grafos SA Arte Sobre Papel, Barcelona
02 03 04 05 06 07 GF 9 8 7 6 5 4 3 2 1

CONTENTS

Plates section appears between pages 82–83.

PREFACE

The opinions expressed in this book are those of the authors and not necessarily those of any employer or government agency.

"Fact" is a relative term and not an absolute determination as generally perceived; all "facts" in this book are according to the judgement of the authors, and again not necessarily of any employer.

The thrust of this book is transportation disasters and response to them. At a very early stage the authors decided to include in this book numerous relevant examples from general disasters, rather than artificially limiting experience exclusively to transportation. Experience is interrelated. It is foolhardy not to gain experience and learn where possible. Case histories, whatever the incident, are an excellent way of learning (Kletz, 1996a). In the same vein, one cannot talk only about the forensic, insurance, or mitigation aspects of a disaster. It is much more accurate to gain an overall understanding of what happens in a disaster and various forces respond. When a response is properly organized, the numerous responders function as in a unified effort. It is not possible to talk about one unit without talking about its contribution to other units and the support that it receives from yet others.

The book is aimed at the responder, who must understand transportation disasters, and at the transportation industry, which must be acquainted with disaster response. Response can best succeed only on the basis of mutual understanding and cooperation. No one program has universal applicability. No one program can be used by every responding organization or involved group. Hence, no specific program outline is suggested; instead, the authors raise personal experiences and points for thought that can be adapted to user needs.

ACKNOWLEDGEMENTS

The authors wish to express their thanks and appreciation to Dr Assa Marbach (Division of Identification and Forensic Science), Esther Tannenbaum (Bomb Disposal), Tzippy Kahana (DIFS assigned to Greenberg National Center for Forensic Medicine – Tel Aviv-Yafo), and Mordechai Cohen (Medical Services) of the Israel Police (Jerusalem); Meier Kaplan and Joseph Almog, Israel Police (retired); Leibel Kruger (Bnei Brak, Israel); James Lederman (Motza, Jerusalem, Israel); Eric Martin of CONCAWE (Bruxelles); John Jay College of Criminal Justice Library (New York); Israel Institute of Petroleum and Energy Library (Tel Aviv); Barry Zalma, Esq. (Los Angeles); Robert Donahue of Logan Airport Fire Department (Boston); Alan Wood, Vice President – Prairies/NWT Insurance Bureau of Canada (Edmonton); W.P. Wilkinson of Seattle-Tacoma International Airport; Rich Hendrickson, Sahuarita Police Department (Arizona); Aharon ("Erol") Eshel, Tel Aviv-Yafo Fire Department (retired); Roar Gronnerud, Norway Police and formerly attached to Interpol, Lyon, France; Derek Stavrou of British Airways at Ben Gurion Airport (TLV) (Lod, Israel); Helmut Vykoukal, Lufthansa Airlines, Frankfurt, Germany; and Shmuel Yarom and Amnon Ilani of Emergency Services, Jerusalem (Israel) Municipality.

INTRODUCTION

ON DISASTERS

There is no clear and universally accepted definition of when an accident becomes a *disaster*. One commonly used definition is "an event causing death and/or extensive property damage, which overrides usual response capabilities."[1] Thus, a road accident with ten injured and ten dead might be called a disaster in a small rural area, but it could well be handled without special deployment and remain an accident in a major urban center.

Another definition is "an event involving injury, death or damage that constitutes a significant disruption to public life." According to this definition the 28 November 1979 crash of a New Zealand airplane on Mt Erebus (active volcanic mountain and "tourist attraction" at 3794 meters high located on Ross Island, Antarctica),[2] with 257 fatalities including 204 complete bodies and 139 body parts, would be termed an accident, not a disaster, since the course of public life was not disrupted. Even when the bodies were returned to New Zealand, there was no significant disruption to the public.

Not every disaster is an *emergency*. This term implies immediate action. At Mt Erebus, once it was determined that there were no survivors, the air crash certainly was not an emergency. There was no one to rescue. Likewise, there is no uniform definition of what constitutes a *disaster site*. Responding agencies have different interests, and they define the disaster site according to those concerns. The importance of defining the scene is more than theoretical, since it is that determination which sets the boundaries of the cordoned-off area and the area of response activity.[3]

Routinely the police are in full control of a crime scene, and it is sole police judgement to cordon off the area; for police the "scene" includes all the zones from which physical evidence is to be collected. In medical terms, the core "scene" is that area where injured or deceased persons are to be found.

The distinction between a disaster and an accident is subjective. In 1998 in the United States more than 94% of 43,020 transportation deaths were road related, yet very few instances could be termed "disasters," even though in many cases highways were closed (constituting a "disruption").

[1] It is not infrequent that the very opposite takes place in certain incidents. For example, ambulances can inundate an area and overwhelmingly outnumber injured. This does not mean that roads can handle traffic, nor that hospitals can comfortably absorb all patients. If these incidents are not disasters, then until their true scope is clarified, they involve disaster-type deployment.

[2] Named after the exploratory ship *Erebus*, active in Antarctic waters in the 1830s–1840s under the direction of figures such as Sir James Clark Ross (after whom the island was named) and Sir John Franklin. The word, Erebus, is derived from Greek mythology (a version of purgatory).

[3] Different functions have different definitions of the disaster area. Fire is interested in a relatively small area. Psychological assistance personnel often work in expanding concentric bands that encompass a greater distance from the incident.

In 1998 no passengers were killed in the United States on scheduled aircraft. In the preceding year 976 perished in aviation accidents, most notably 228 persons who died in the Korean Air Flight 801 in Agana, Guam (a US possession included in US government statistics). Only the larger aviation accidents are called disasters; smaller aircraft which crash are again considered accidents[4] (sometimes relegated to unnoticed items in the inside pages of the newspaper).[5]

[4] Unless they crash on populated areas.

[5] In Israel most aviation accidents involve crop-dusting planes that hardly have the potential to be called major disasters.

Of the 746 US rail-related deaths in 1997, 584 victims were persons walking on or near tracks. Thus, the vast majority of these rail-related deaths are also not considered "disasters." (Canadian statistics show a similar danger from trespassing on the rails.)

On the other hand the International Red Cross Committee recorded 700 disasters in 1999, essentially occurring at the rate of two a week. On the one hand these statistics include the Third World, where even major disasters are essentially ignored by the developed world; on the other hand this large number can be considered a tacit validation of the Red Cross's existence and function. Admittedly, however, the Red Cross does not (and cannot) offer assistance in every instance.

Another factor influencing the use of the term "disaster" is the very subjective evaluation of people who are quoted and of the media. If the "right" people (such as a television news anchorman) call an event a "disaster," there is a better than average possibility that the term will catch on. Popular use of the term "disaster" will also make it much harder for government to avoid officially designating the incident as a disaster.

Many of the examples cited in this book are not disasters. In absolute terms they are accidents. These examples were chosen because they highlight issues that also happen in disasters.

ON RESPONSE

Disaster response was once a straightforward procedure of dispatching medical, fire and police personnel to a scene to do whatever they could – more or less depending upon the situation. We also sought simple explanations of why disasters occur.

Today, disaster response is much more sophisticated. We know that many disasters are not caused by a single factor. We have designed numerous safeguards (against hurricanes there are weather-related precautions, building codes, etc.). A disaster is not a single failure. It is a system failure (Taft and Reynolds, 1994). Disaster response must, therefore, address a system and not a single factor.

Disaster response is, in fact, only one aspect of an overall program of emergency management. As this book will show, *comprehensive emergency management* is both a philosophy and a profession designed to maximize benefit for all involved. It uses skills in management science, psychology, medicine, and other fields of knowledge to produce a coordinated project to serve disaster victims.

Instinctively, one might think that it is of little matter whether an incident is called a "disaster"[6] or an "accident." Response to the occurrence is based upon need. When the fire department gets a call, they do not decide if they have the budget to go. They simply respond with any forces that are required. The descriptor used to classify the incident is of little concern.

[6] Etymology: an event under the negative influence of the stars.

A good part of disaster response does require immediate response to save lives and guard public safety, but there is much more to the issue. Definition of an incident as a disaster is often the key to the release of supplementary funding and even government agency empowerment. Very simply, there are often steps that can be taken after an incident has been confirmed as a disaster. This is quite blatant in the United States with the release of Federal Emergency Management Agency (FEMA) funding only after a presidential declaration of disaster. This funding then allows the implementation of certain government programs, such as relief distributed under the Stafford Act (PL 93-288 as amended).

With all that said, it should be remembered that disaster is a heavily emotional experience. It is not unusual that disaster procedures are initiated in an incident simply because the word "disaster" has been used in popular speech, even though no government agency has made an official declaration. One of the biggest challenges in disaster response is to standardize definitions and procedures to enable a more orderly response mechanism. More and more disasters are becoming global incidents involving multi-national victim populations. This means that a new perspective must be recognized, discarding localism and replacing it with internationalism.

ON THOSE INVOLVED

There are more people involved in a disaster than might first be thought. Taft and Reynolds (1994) developed a function-oriented list of organizational involvement in a disaster. This list can easily be adapted to personal involvement.

- *Primary:* major victim, such as a railroad or bus
- *Auxiliary:* in fire, aircraft manufacturer
- *Alleviating:* direct responders (fire, police, etc.)
- *Unionate:* trade unions involved
- *Pressure:* such as families
- *Commissioning:* boards of inquiry

To this list one can add secondary responders such as hospitals, with further dichotomy as treatment continues over the long term.

A key element in disaster response is isomorphism, the opportunity for one organization to learn from the experiences of another organization (Taft and Reynolds, 1994). All airlines can learn from an air disaster. Airport authorities can learn from carriers. Aircraft manufacturers can learn from air crashes. That is what this book is all about – one person or organization learning from the experiences of others. But not only learning! All lessons learnt must be integrated into disaster planning if they are to have any practical benefit.

Isomorphism means learning from all possible sources. If a transportation company can learn a lesson from a hotel fire, it certainly should. For that reason, this book includes examples from relevant non-transportation disasters and accidents.

ON THE FUTURE

Disasters are negative events. They destroy. They injure. They kill. But they can also be positive motivating events that become milestones of attitudinal change and force a new future.[7] The Great Chicago Fire was the precursor to new programs of urban planning. The sinking of the *Titanic* ushered in new concerns for maritime safety. The explosion of the *Hindenberg* closed the era of the dirigible. Lockerbie ushered in new concerns for aviation security. The Swissair crash signaled greater acceptance of emergency landings, even when the cause is not *verified* to be life-threatening. Sometimes it takes an extreme event to shake people and institute change. The best time to institute that change is immediately after a disaster, when the experience is still fresh in the mind (Stevenson and Hayman, 1981).

[7] For a discussion of whether disasters cause *social* change, see Morrow and Peacock (1997).

SECTION I

TRANSPORTATION AND TRANSPORTATION DISASTERS

Chapter 1
BACKGROUND FOR NEW DISASTERS

CHAPTER 1

BACKGROUND FOR NEW DISASTERS

INTRODUCTION

Since the dawn of history humans have exhibited an insatiable curiosity regarding their surroundings; this has led them to undertake incessant and restless wanderings away from their earliest settlements. Inevitably, these adventures were at first undertaken on foot.

Four modes are generally identified in modern transportation: water, road, rail, and air.[1] Each of these has developed in its own way over the centuries, and its stages of development have had important effects on the advantages and disadvantages of the alternative forms of transportation available at the time.

Invention of the wheel by our early forebears accounts for one of the great breakthroughs of human technology. Even so, transportation – the conveyance of persons or goods – was long powered by animate sources of energy. Animate power for land transport usually curtailed the weight of what could be moved and rarely allowed speeds in excess of 16 km/h.

[1] The thrust of this book is the transportation of people. Trucks carrying cargo are included because they use the same highways. There is a driver and sometimes a passenger. Cargo aircraft are also considered, again since there is a crew. The response to a "cargo only" flight can be very similar to a "passengers and cargo" flight. Pipelines, however, are not considered in this book, since they transfer liquids and do not transport people.

WATER

Bodies of water provided defense, convenient waste removal, food and drink, and a means of getting from place to place quickly. Human settlement was frequently on the shores of rivers, lakes and seas. It was on water that humans first escaped from the constraints of live sources of power. They first learned to exploit the existing currents of rapid streams, rivers and seas, eventually building ships, which harnessed prevalent winds for sails to supplement or replace their oars. Canal building flourished in the early 19th century until growing rail networks proved more efficient and economically viable in most industrial countries.

Once steam had been harnessed to drive the budding industrial revolution, application to shipping was an obvious step. Fulton's steamboat, the *Clermont*, made its maiden voyage in 1807. The diesel engine has since replaced steam as the primary source of power for the world's ships, while enhanced designs such as stronger hulls, hydrofoils and hovercraft have facilitated greater capacities and faster ships. In recent decades it appeared that shipping would be relegated to freight alone since sea-lanes could not compete with the speed and convenience of air travel. Now, growing affluence and increased leisure in the world's wealthier countries seem to have given new life to luxury cruise liners, some taking on the size and diversity of floating cities.

RAIL

The industrial revolution depended on the movement of raw materials from their source of origin to the regions of industrial activity. Vast quantities of fossil fuel, first coal and then oil, were required to run industrial machinery. End products had to be supplied to markets around the world. Hinterland areas, ever further distant from industrial cities, had to be made accessible to provide and feed a growing urban labor force. While sea-lanes, then as now, could provide for many early intercontinental requirements of colonial and industrial empires, canals and inland waterways soon proved inadequate for the internal logistic needs of industrial nations and continents.

The solution lay in technological improvements in rail transportation. The idea of rolling carriages over wooden tracks, then metal-covered tramways and finally metal rails, evolved from the 17th century. Adapting steam power to an engine for locomotion replaced the need for animals to power these operations. James Watt designed the first efficient steam engine in 1769. Stephenson's locomotive pulled the first train of cars in 1814, and 1825 saw the first regular train service inaugurated in both Great Britain and the US. Soon, vast networks of rails traversed the continents and provided the means to reach into hitherto inaccessible parts of industrializing nations to move people and goods. Changes in fuel from steam to diesel and electric power have made rail transport more environmentally friendly. Their dominance was unchallenged until the advent of the motor vehicle. Despite trucking, railroads are still unsurpassed in moving bulk goods, but their share of passengers has declined in the face of automobile and air travel. Only light rail systems and subways still compete for passengers in crowded metropolitan environments, even when they are hard pressed financially. New technologies suggest many innovations in the foreseeable future.

ROAD

Development of road networks actually preceded the age of the motor vehicle by as much as a century. Existing country roads were often expanded and improved to facilitate population movement, transportation of goods and access to hinterlands. Since the introduction of the mass-produced automobile, its flexibility has proven to be such a great attraction that no alternative form of transportation has thus far been able to compete with the passenger car.

Dominance of the automobile has actually turned into something of a mixed blessing. It may be that the performance capacities of the modern automobile exceed road conditions and human capabilities, and new technologies will have to be applied to assist in managing the automobile of the future. As cars have grown more numerous and powerful, concern over traffic accidents, congestion and pollution has increased. These issues are now finally being addressed both in governments and industry circles.

AIR TRANSPORTATION

Air travel is a 20th-century phenomenon. More than other forms of transportation, commercial air travel has benefited from war-related technology in the wake of the First and Second World Wars. Since the Second World War, increased speed and safety have attracted millions of passengers, lowering fares and making air travel even more attractive to large sectors of modern society. Since the greatest inconveniences of air travel involve delays caused by overcrowded flight airspace around airports and extended pre-flight and post-flight processing, air transport is most efficient in facilitating travel between long distances.

MATERIALS FAILURE AND HUMAN ERROR

Since travel in more affluent societies is unavoidable and even part of our increasing leisure, we would all want to be assured that our means of transportation are always safe. Unfortunately, this is all but impossible. Nevertheless, great effort has been invested in trying to reduce the chances of mishap.

In today's sophisticated means of transportation the sudden failure of the smallest sensitive component could spell disaster. Technological advances were first reflected in this area by the introduction of stronger, more wear-resistant materials and designs. Nevertheless, all equipment tends to wear out with use so that routine maintenance and careful monitoring are necessary.

The quest for quality assurance was introduced under the influence of the insurance industry. Preventive replacement of sensitive parts before their reliable life span approached its end was determined by their estimated life expectancy. Regrettably, this is a prohibitive procedure not universally practiced.

TRENDS

There has been a logarithmic increase in the movement of goods and passengers since the Second World War. Architect Buckminster Fuller, inventor of the geodetic dome, once said that his father lived his life in the space from one horizon to the other, while he clocked a hundred thousand miles of annual worldwide travel. The products in our markets are also less likely than ever before to be local products. These realities testify to the changes in transportation taking place before our very eyes.

In recent decades, growing attention has focused on the role of human error in both industrial and transportation disasters. Determining the place of human error in disaster is not always simple. Equipment failure during operations may arise from poor development and design, which can rightly be laid at the doorstep of the planners and engineers. It could also result from faulty parts, unauthorized substitutions or deficient maintenance practices – all human failings.

Some individuals may be mismatched for the type of work they do, but all individuals are subject to human error. The problem is so pervasive that equipment needs to be planned with typical human lapses in mind. There are also typical errors in decision making, both at the individual and the group level, that require close attention if they are to be avoided. Safety engineers and reliability analysts typically pay closer attention to equipment failure than to background social and psychological factors that predispose to error (Collins and Leathley, 1995; Granot, 1998).

Attitudes toward disasters have undergone change over the twentieth century. Where once traditional societies accepted such unpredictable incidents with fatalistic resignation, modern societies began making contingency plans to reduce losses. Finally, efforts were turned to prevention. Chaos theory studies seeming disorder in dynamic systems and catastrophe theory uses sophisticated topology to explain the unanticipated association of unrelated items to explain a disastrous sequence of events. Both scientific theories lend credence to the view that mitigation is possible but prevention is probably not.

Regardless of the specific means of transportation being considered, the overriding considerations are speed, economy and increased capacity. There is constant pressure to move larger payloads, faster and cheaper. While this does encourage technological innovation, it also provides an incentive to cut corners and take risks. Modern means of transportation are no doubt immeasurably more reliable than their predecessors, but their intricate electronic systems may actually be less robust than mechanical systems under human control. When they do fail, their increased capacity may expose larger numbers of persons to risks greater than ever before. As a result, as natural hazards have declined in modern societies as a threat to life and limb, transportation-related mishaps have become an increasing danger.

SPECIAL CHARACTERISTICS

Mass disasters tend to have many common features. Today more is known than in the past about their causes. Disasters tend to progress through stages. Much has been learned regarding the typical ways in which individuals and societies respond in these unexpected circumstances.

Despite the features shared with all disasters, transportation incidents do have their own special features. Regardless of the means of transportation involved, the victims using any means of conveyance may be very distant from their homes. In many instances they are chance travelers, unknown even to one another. Emergency response depends on the chance location of the incident. Unlike the sympathy usually evoked when neighbors suffer, those living in the vicinity of the accident may not identify with the chance victims in their midst. There have been recorded incidents of resentment and even hostility toward outside travelers who fall victim in a chance location. In one incident where a plane crashed near an isolated community in Great Britain, thieves even took the trouble to leave the graffiti message, "Better Luck Next Time!" Regardless of these spontaneous feelings, local resources must be allocated to cope with the emergency that may just coincidentally land on a community's doorstep.

On the other hand, a distant incident causes great hardship to services and families of the dead and injured. That distance can increase the anguish and the financial burdens of all concerned.

CONCLUSION

When something goes wrong with our modern means of transportation there is fairly widespread empathy and concern on the part of the general public. Few persons today can avoid the use of a variety of means of transportation. A failure of any sort elicits concerns over the possibility that the same could happen again. Modern transportation involves relatively sophisticated equipment usually operated by trained personnel. There is always a strong need to determine whether the cause of an incident was in equipment, human failure or some combination. A work that examines transportation disasters can contribute significantly to greater understanding in this sensitive area.

SECTION II

GENERAL ASPECTS OF DISASTER RESPONSE

CHAPTER 2

HISTORY OF DISASTER PLANNING

INTRODUCTION

Transportation tends not to have specific disaster response personnel except in the aviation industry, and then primarily in airports and in the headquarter offices of a very few airlines. In most airlines, *if* there is a disaster response officer (not to be confused with a flight safety or occupational safety officer), that appointment is most often part-time, if not totally symbolic. Thus, most transportation disasters are handled within the context of general disaster planning.

CIVIL DEFENSE BACKGROUND

As a historical note it should be mentioned that much of disaster planning had its roots in civil defense programs developed before and during the Second World War, and during the Cold War (Dynes, 1994). These plans totally disregard civilian and non-military needs (such as public transportation). The plans are based upon the notion that a rigid hierarchical command system is needed to handle disasters, as the military functions in war scenarios.[1] This is implicit in the US Federal Civil Defense Act of 1950. Mileti (1999) suggests that modern disaster research began during this same period of military-oriented planning.

In the United States as late as the 1990s it was necessary to have such programming in place for disaster funding to be released. Only in relatively recent years has planning been shifted to civil protection, stressing non-military accidents and their response (albeit still cast in a military mindset). At this writing there is a movement in the United Kingdom to replace provisions of the Civil Defence Act 1948 with an act stressing civil protection (ICDDS, 2000). The British Institute for Civil Defence and Disaster Studies is now emphasizing civil protection; as their name suggests, their beginnings were in the pre-Second World War atmosphere of aerial bombings and acts of war (Jackson, 1988).

Many of the remnants of civil defense can still be discerned. In many areas disaster shelters are still planned on the model of the 1950s fallout shelters. In

[1]It can be inferred from Dynes (1994) that turning over disaster response planning to police and fire services, both hierarchical organizations, is a remnant of the military focus and mindset.

[21]Hebrew, PeSaCH.

Israel, for example, military thinking in civilian offices is particularly pronounced. The Israel Ministry of Interior disaster planning office[2] is still staffed by retired military careerists. Not only do they bring their military frame of mind with them; they receive relatively little training to handle civilian problems (except for on-the-job training most often provided by other military retirees). Although there was an Israeli exercise regarding an incident involving a train during the early 1990s, the scenario went along the lines of military thinking and not according to a course-of-business civilian accident.

This same military approach was true in many parts of the United States for the career span of Second World War veteran soldiers (in many cases well into the 1970s). This entire approach stressing the war perspective and shelters left little place for response to major transportation accidents (of which there were quite many). Gerber, writing in 1952, summed up emphasis in the military-oriented civil defense approach, "The Civil Defense set-up affords opportunity to develop an organization prepared to meet any emergency, not just those resulting from enemy attack or sabotage."

HIERARCHY

The idea of hierarchy is a concept well cherished in the military. In civilian planning it has a problem. It assumes that upper echelons know what is best, and they impart their wisdom downward in the system (Dynes, 1994). In reality, upper management in civil administration has little real disaster experience (except for a prior incident that might have happened during their tenure); but, senior echelons *do* have management skills. Disaster-specific knowledge, both theoretical and practical, is more often found at the lower, technical levels of bureaucracy, where workers tend not to have broad-based management skills.

MILITARY MODELS

Due to the close relationship of civilian disaster response and military civil defense, armies have applied lessons learnt in disaster response to operational problems found in the army. Yet, there are vast differences between civilian accidents and military events. One cannot compare a civilian air crash with passengers from numerous countries, to a military flight with a known passenger list and in virtually all cases medical records already on file for identification purposes.

The military also operates on coordination by legal mandate; there is no civilian counterpart. International cooperation operates on a basis of good will. One must also remember that wartime thinking is based upon different priorities, different capabilities, and a different legal structure – even in the civilian sector. The affected population is different in times of war, since it is plausible to conjecture that most able-bodied men will be in uniform or on their way to report for military service. These factors definitely affect the decision-making process and highlight once again the inappropriateness of forcing military models on civilian applications.

As discussed, the roots of *government* planning are in the military experience. However, the roots of *private business* sector disaster planning are in security programs (whether for plant security, compliance with government regulations, or insurance company dictates). In recent years in the private business sector this planning has taken a new identity upon itself as *business recovery*. This is not a new subject, but it is only lately that it has been formalized.

An early example of business recovery is the 30 July 1940 explosion and fire at the R.H. Hollingshead Company in Camden, New Jersey (USA), causing extensive damage to the area telephone system. Within less than an hour PBX trunk lines were set up elsewhere in the city as part of a pre-planned telephone network recovery program.

SECRECY

Disaster planning is not inherently "secret." This is another remnant from the era of military-based civil defense planning. There are, of course, certain disaster plans that cannot be released to the public, but they are few and far between. The classification stamp should be used sparingly and with adequate forethought.

Public discussion is, in fact, a positive aspect of disaster planning. Risk assessment and examination of contingency plans at open meetings allow for positive input and raises general awareness. This is important, since the biggest enemies of disaster planning are the intertwined phenomena of public apathy and the downplaying of hazards.

CONCLUSION

The historical background of disaster planning consists of military models and civil defense planning. The challenge now is to civilianize thinking in all respects and tailor disaster management to new concepts. Part of that new thinking means an increased participation of elected government.

CHAPTER 3

THE ROLE OF GOVERNMENT IN DISASTER PLANNING

INTRODUCTION

When military thinking dominated disaster planning, *command and control* were central. Decisions were filtered down from commander to the field. Civilianization means new concepts. One concept is cooperative understanding.

The role of government must be understood by all those involved in disaster planning, including the private sector. Likewise, government must understand the broadest needs of the effected area – both residents and businesses. This did not necessarily exist in the era of command and control, when dialog was in one direction only.

GOVERNMENT

Numerous layers of government try to provide services to modern society. To cite the typical example of an American town, the local municipality is at the bottom of the totem pole beneath layers of federal, state, and county administration. In England, Canada or Australia there is little difference. Even in Israel, where the relatively small size of the country promotes centralization, there is a national layer of government, then regional and local layers.

It is ironic that the greatest burden of management filters down to the smallest element – local municipal government, or regional government when talking in terms of an unincorporated area. If this is true in the functions of daily government, it is all the more true in times of disaster. Even when upward calls are made for assistance, responsibility remains with the local municipality. This occurs even in a case seemingly unrelated to the normal functions of the jurisdiction, such as the crash of a bus passing by on a highway cutting through the jurisdiction. An extreme example is a plane overflying a country in which it does have a scheduled landing. Then the local municipality is drawn not only into the disaster, but into the governmental foreign relations hierarchy with which it has no real experience.

The British Home Office (n.d.) is very direct in assigning this responsibility to local government: "It is fundamental to the arrangements for dealing with disasters in the United Kingdom that the first response to any incident is at the local level." Their solution for disasters beyond the capability of local authorities is the activation of pre-planned mutual-aid pacts.

In the United States this concept of local jurisdiction now holds out the potential of prosecution when a local authority is perceived not to have discharged its responsibility properly (auf der Heide, 1989).

At every level of government, disaster response responsibility must be agency-specific with clear definitions of what is to be done. Otherwise a tennis-like situation is created in which one agency can try to volley responsibility into another agency's court. If responsibility is a local task, then coordination is also a local problem. In any disaster there are a large number of "players," whose work must be brought together into one coordinated response. The difficulty is that disaster response introduces new elements into a local atmosphere. Every locality learns to deal with house fires and car accidents, and to coordinate the activities of responders. Disaster response, however, is not merely a "larger fire" or a "more serious accident." Disasters mean the inclusion of response agencies and individuals who are not involved in smaller incidents.

It is ironic that in many government plans, not necessarily restricted to disaster, the private sector is excluded. When private companies are included, it is to provide specified services on the basis of a legal contract, with overall responsibility remaining in the hands of government.

RESPONSE PROGRAMS

A response program must be of wide perspective. The fact that in a particular situation residences have not been destroyed displacing inhabitants does not mean that there are not other housing considerations that must be included in the response. Very often after a train or airplane crash, for example, uninjured out-of-area victims must be housed. Accommodation must be found for visiting relatives. The non-local press and responders from other cities (e.g. representatives of national government) must also find housing. Sometimes that housing can be for a week or more. All of these requirements go well beyond the response to routine accidents.

In everyday life government agencies work more or less independently, each doing its own task with a minimum of inter-agency coordination (Granot, 1999). In a disaster that is certainly not the case. It is insufficient to have a series of independent response programs to provide health and housing assistance. Programs must be coordinated, and that is the job of local government. Even when the local authority itself provides services, a strong measure of coordination is necessary. In an air crash in New York,[1] for example, the city government

[1] In the United States the Federal Emergency Management Agency (FEMA, established in 1979, based upon predecessor agencies dating back to the 1933–1939 National Emergency Council and its successors) generally does not play a role in transportation disasters. There are, however, roles played by the National Transportation Safety Board (NTSB) and the Federal Aviation Authority (FAA). Those roles tend to be limited to the functioning of aircraft and cabin-crew/flight controllers; they do not investigate matters of post-crash response.

had to coordinate its response services with more than one hundred different outside entities.[2]

Designing an effective disaster program is not simple. In most instances government is reactive. In disaster terminology that means that most governments react to disasters once they have happened (the response and recovery stages), rather than deal with pre-disaster preparation and mitigation. Preparations get bogged down in the budgetary process before a disaster (Lindell and Perry, 1992). This is true particularly regarding transportation disasters, since they have no local focus (except for an airport, which is most often outside the municipality). Response programs do have at least one benefit. There is much more bureaucratic financial flexibility once a disaster has taken place.

Another aspect of being reactive is that government plans disaster response programs according to its perception of *public demand.* It also sets the level of its response (both in size and finance) accordingly. One type of *response failure* is when the bureaucracy continues to provide service at one particular level without perceiving a change in public attitude.

To cite one example, internationally there has been increased interest in disaster victim identification since approximately 1980. Personal recognition has yielded to scientific methods. What was acceptable in 1975 was to be criticized only ten years later. (In Israel the police moved along with the changed attitude towards science; the army kept stressing personal recognition in victim identifications. The net result was the police's usurping of the traditional Israel Army role in identifying even civilian dead.)

[2] Scanlon has used the term "players."

LOCAL RESPONSIBILITY

There are numerous models to illustrate local responsibility. In many cities municipal responsibility during a disaster is defined as the provision of routine municipal services in non-routine situations. Life-saving activity is assigned to police, fire (in fact, municipal, but operated independently), and ambulance. The municipality deals with welfare (emergency housing, food, clothing; social services; psychological assistance), sanitation (clearing debris), quality of life (replacement of damaged road surfaces and signs, utility poles), etc. Obviously, the municipality makes available to other response services all other capabilities that might be needed, such as trucks and mobile lights.

Jerusalem (Israel) has suffered numerous terrorist attacks (some of which are subjectively called disasters in common local idiom), hence the municipality has developed its own philosophy and infrastructure to deal with these events. Many other cities find themselves unprepared to cope with their "first"[3] event that is considered to be a disaster.

[3] "First" in institutional memory.

Since the municipality is not involved in the immediate requirement of saving lives, experience has shown that it is best to limit the number of its responders in the first hour of a disaster response. Large numbers of municipal responders constitutes *professional convergence* and only interferes with the response.

A main function of municipalities is the operation of an Information Bureau,[4] described below.

[4] A complex question is the operation of an Information Bureau when "nothing happened," i.e. disaster was avoided. This came into play when a car bomb was detected before explosion in a Jerusalem neighbourhood in March 2001. The response – controlled detonation – closed numerous main streets for several hours. Since there was no disaster, it was decided not to open an Information Bureau. Information was released through routine city offices.

EMERGENCY MANAGER

In a proper arrangement a local government will appoint an emergency manager, full-time or less depending upon the size of the municipality, to plan for disaster issues. Those plans should take into consideration the widest range of possible problems, such as gas leak, Hazmat (hazardous material), fire – and transportation. Planning must be innovative, since hazards do not necessarily occur in expected areas. Hazmat, for example, is often carried by winds from spills or accidents in an industrial plant, a pipeline, or a truck. The Hazmat danger may well exist in a jurisdiction other than the one in which the initial incident took place. Not only do winds carry Hazmat and pollutants; there are all-too-frequent polluting spills from cargo ships and tankers.

All of these incidents can endanger life in populated areas. All of these incidents are best handled by in-place emergency plans. In many cases a local government does not have an emergency manager. The municipality has to draft workers from other municipal government jobs. Even if there is an emergency manager, he often does not have a large staff. A common mistake is that all too often a relatively low-level clerk is given the "unimportant" job of emergency planning (unimportant along the lines of wishful thinking that a disaster will not occur).

The 31 January 2000 spill of cyanide into the Tisza River shows why plans should be flexible and well prepared. Most damage was done in Hungary and Serbia, although the spill occurred in Baia Mare, Romania. The death toll was in river life, most notably fish. For area residents there was a danger only when using waters from the river, either directly (drinking) or indirectly (e.g. eating fish from the waters).

In fact, the emergency manager must have a broad perspective, seeing disaster planning, response, and recovery with a wide view. Plans should start as a natural by-product of routine work, so that a skeleton crew is in place when a disaster does occur. The emergency officer must also have the stature and authority to activate both additional personnel and financial resources when a disaster happens. When doing so he must not be confronted by a series of agencies, each working under its own set of internal rules. He must envisage a large and coordinated operation in which numerous entities work toward a common goal. People must be given assignments in technical areas. There must be, however, someone with overall responsibility, who makes sure that all technical functions work in unison with each other. That is the job of a capable emergency manager.

A basic rule is that whoever deals with a problem will most often perceive it from his routine point of view, and his proposed solution will be from that perspective. For this reason the manager must appreciate the overall picture and see the broadest implications of a situation. If a person in charge of sanitation is placed in charge of a flood, he can be expected to stress drainage problems. When a traffic engineer views that same flood, he is more apt to analyze questions related to the movement of emergency vehicles. Both approaches are correct, but both must be looked at in the overall context of a disaster response.

A transportation company is in a particularly difficult position regarding local jurisdictions and its influence on their response planning, even though that company is a "potential client" of municipal services. An example could be a long-distance bus company whose route passes through a series of local jurisdictions (very often not even picking up passengers or even stopping at a rest station). As long as there is no accident, the company has only a theoretical interest in local disaster planning. It is also rather difficult for the company to voice requests to the local jurisdiction, since it pays no local taxes nor do its employees vote in local elections. In many cases the bus company is virtually unknown, since its buses that bypass the community have absolutely no effect on community life and are not even felt, except as an anonymous traffic statistic – until there is an accident.

In aviation the situation is slightly different, but not much. Airlines do have an address where they can voice disaster planning concerns – the local airport. In many cases, however, planning that extends outside the airport perimeter fence is superficial at best, even in terms of close-by localities under standard flight paths. An airline has no contact with those jurisdictions not in immediate proximity to the airport (such as the scores of jurisdictions over which a flight from New York passes from the coast of Europe until landing in Frankfurt or Vienna).

In contrast, transportation company/local authority interaction is quite comprehensive after an accident. In the case of an air crash into a built-up area (e.g. Air France Concorde near Charles de Gaulle Airport, El Al in Amsterdam, SavanAir in Casenga, Angola or Air Caraibes in the Netherlands Antilles), a comprehensive plan would include not only fire, police and medical personnel, but also possible responses from the tourism board, psychological services, local transportation agencies, communications, welfare (to provide clothing), public utilities, perhaps an agency dealing with foreign language translation, etc.

On 2 February 1999 a SavanAir (Antonov 12) cargo flight crashed after takeoff from Luanda. The plane hit houses in the suburb of Cazenga. Reports described 20 fatalities on the ground and at least two houses totally destroyed.

On 24 March 2001 at about 1630 Air Caraibes Flight No. 1501 (DHC-6 Twin Otter, code-shared with Winair), crashed into residential structures on a flight from the Caribbean islands of St Maarten to St Barthelemy. All 17 passengers and crew as well as a man on the ground were killed; a woman on the ground was seriously injured.

Air Caraïbes is an area carrier that combined parts of Air St Martin, Air St Barth, Air Guadeloupe, and Martinique to serve the eastern Caribbean region.

The Air Caraibes flight posed a common response problem. It was decided to bring in French experts to identify badly burnt victims, both in view of the lack of local experts and because there were 12 French (including one dual[5] US/French) citizens aboard.[6] Both Air Caraibes and its code-sharing partner shared aviation related-response functions (e.g. dealing with bereaved families).

[5] When examining passenger lists, the possibility of dual citizenship should always be taken into consideration.

[6] Other passengers were from the Netherlands (1), Belgium (2), USA (1), and Monaco (1).

PRIVATE SECTOR – GOVERNMENT INTERDEPENDENCE

In usual circumstances private companies provide services. When the private sector (e.g. a transportation provider) is a direct victim of a disaster and cannot provide service, it is the government that holds the responsibility to assure that basic services be provided. If a company immobilized by a disaster should have provided a response service, it is the responsibility of government to find an alternative provider.

CONCLUSION

Modern approaches dictate that government is a partner, not a commander, in disaster management. That partnership includes local citizens and private businesses including transportation companies. They must all work together to plan for their mutual well-being. Not to be forgotten, of course, is the well-being of disaster victims.

CHAPTER 4

VICTIMOLOGY

INTRODUCTION

It is important to know the population for which disaster plans are prepared. Not all populations are equally vulnerable to disaster.

NOT EVERYONE IS EQUALLY VULNERABLE

Wijkman and Timberlake (1984) show that in natural disasters, as national income increases, fatalities decrease. Another researcher takes this one step further and estimates that over 90% of disaster-related deaths occur in developing countries. A conclusion needing further study is that transportation disasters are more survivable in developed areas; the same disasters cause more fatalities in non-developed areas. It would seem that major reasons for better performance in developed countries are both in improved mitigation and better response (particularly in the medical and fire-fighting areas).

To cite one example, an airline official described the emergency landing of an aircraft in one developing country's major airport. A small fire became a major conflagration, after hoses from the airfield fire department leaked and could not maintain water pressure. The airline then permanently suspended all flights to that developing country.

Poorer populations find inexpensive housing, often in disaster-prone areas. This includes areas such as, for example, floodplains. The same holds true, though to a lesser degree, in transportation disasters. It is the poorer segment of society that lives in the "noise areas" closest to major highways and train tracks, hence it is they who are the first victims of pollution and Hazmat spills. It is often those same poor who live close to airports and under ascent/descent flight paths, hence it is their houses that have a higher chance of being hit in takeoff/landing accidents.

Regarding aircraft, accidents involve primarily those members of society who use air transportation (with the rare exception of those few[1] unfortunate enough to find themselves in the ground area where an airplane crashes).

Granot and Levinson (2001) point out that as urban terrorism turns to acts

[1] For example, on Sunday 3 September 1989 a Cubana Airlines jet en route from Havana to Milan crashed shortly after takeoff. At least 35 people on the ground were killed and "some 30 houses . . . were completely destroyed." (Granma, 10 September 1989, p. 11; also Bohemia, 8 September 1989, pp. 32–33.)

against public transportation, the higher-income car-oriented segment of society is less likely to suffer casualties.

KNOWING THE POPULATION

A clear demographic picture of the population is a precondition to a sound disaster plan. This is most blatant in a transportation disaster such as the leaking of Hazmat during transport and subsequent evacuation. Evacuating a residential area is different from evacuating an industrial zone. Little children have special needs, even in a short-term evacuation, ranging from diapers and formula to activities to keep them occupied. Deaf people might not hear verbal warning messages such as those announced by loudspeaker or siren, or broadcast on radio or television not accompanied by sign language. A senior citizens home can be merely assisted living, or it can be a virtual geriatric hospital (Taft and Reynolds, 1994). Children and the elderly are not found in factories. Handicapped are found in fewer numbers in an industrial zone, unless there is a business with a special program. In all of these cases a detailed knowledge of demographic distribution is a prerequisite to emergency planning (Ferrier, 1999–2000).

SENIOR CITIZENS

Older persons (Baldi, 1974; Poulshock and Cohen, 1975; Hutton, 1976; US DCPA, 1976) are more vulnerable to disasters and to serious medical problems such as cardiac complications; and, they more often need special assistance (transportation or other) in evacuation. In other words, an accurate projection of population details is a critical tool in the eventual provision of appropriate disaster response services. For this reason there must be diversified planning. Plans that are appropriate for one population are not necessarily appropriate for another population. There can also be no "transportation disaster plan." There must be an overall plan as part of a disaster-response frame of mind that encompasses a broad range of possibilities, including transportation.

Hutton (1976) conveys a very interesting research conclusion. It is usually thought that the higher death rate amongst the elderly in disasters is due primarily to their difficulty in receiving warning messages (Friedsam, 1962). According to Hutton this was not the case in her research; there was no significant difference between young and old in receipt of warnings. There were also no major differences in evacuation according to age. Assuming these conclusions, based upon the Rapid City, South Dakota (USA) flood, are correct, one wonders if the results are representative or unique to this event. According to Hershiser and Quarantelli (1976) no real emergency plan was in place at the time of the disaster, and most of the response was based upon personal intuition.

On 9 June 1972 a flood struck Rapid City, South Dakota, USA. 238 people were killed and more than 3000 injured. The flood destroyed 770 houses, 565 mobile homes and 5000 cars and trucks. In total it caused more than $165 million in damage.

VICTIMS ARE NOT HELPLESS

It is totally false that disaster victims cannot cope without outside assistance (Cuny, 1983). Even the most vulnerable population can have a response capability (even if it is overwhelmed by the dimensions of the disaster). Sometimes survivors perform heroic acts to save others. Although this truth has been often repeated in professional literature, the myth of a victim's helplessness continues, perhaps due in part to the nature of relief organizations. As Brown (1995, pp. 25–26) points out, private relief organizations must spend money (the "pressure to spend syndrome") to justify their existence to donors. Thus, there is a vested interest in giving assistance, be it needed or not. One might ask if the same is not somewhat true for governmental international organizations, which also want to demonstrate their presence.

There are numerous reasons for the continued existence of disaster myths, even amongst disaster workers and planners. In part this is due to the fact that only in recent years has disaster management begun to take upon itself the trappings of a learnt discipline with a base of literary experience and analysis. This attitude still has not totally penetrated municipal, police and other response agencies, where disaster management is all too often considered an assignment rather than a profession. With that attitude, workers do not read books or even journal articles – nor are they even aware of specific academic resources. The net result is that except for an occasional disaster course, the basic sources of "professional" knowledge are radio, television, the newspaper, and the man on the street.

Self-rescue is also neglected by the press. This rescue occurs in the first moments after a disaster, before reporters equipped with their cameras arrive on the scene. Reconstructing what happened can be extremely difficult, particularly since these self-rescuers are also victims; once formal forces arrive, the informal rescuers resume their anticipated roles as victims.

Spontaneous responders are also ignored by the media for similar reasons. Their roles terminate upon the arrival of professional disaster responders. The press usually checks with its usual and established sources for news stories.

Spontaneous responders are outside that usual channel (Quarantelli, 1996). Anything beyond isolated self-rescue incidents also challenges the myth that victims are helpless.

DISASTER SUBCULTURE

Granot (1996) surveyed previous sociology literature discussing the existence of a disaster subculture. He concluded that, "It is not an alternative to the mainstream culture of a society. It really represents an aspect of that dominant culture that only manifests itself under particular circumstances."

In terms of victimology, there are disaster subcultures amongst certain segments of society that have previous disaster exposure or experience. During the wave of bus bombings in Israel during the mid-1990s, segments of Israeli society developed their own subculture. There was an attitude of fatalism and an awareness of danger that characterized the daily bus commuters of Jerusalem and Greater Tel Aviv.

VICTIM ASSISTANCE

The victims of disaster are often in need of a broad range of assistance projects to assist their return to normalcy. This can include such aspects as psychological services, housing programs, and legal counseling. Assistance is most often rendered by local government or by private companies involved in the disaster. The latter is particularly difficult for private companies to provide, since they possibly have their own vested legal interests.

In the United States a private non-profit organization, the National Organization for Victim Assistance (NOVA), has defined its role as assisting not only victims of crime, but also those involved in a disaster. It has thus tried to fill a void in victim assistance.

CONCLUSION

Knowing the population is a precursor to designing a response plan to meet needs; however, those needs change as a disaster progresses. In each stage needs are different. Understanding the progression of stages in a disaster is a corollary to victimology.

CHAPTER 5

DISASTER CYCLES

INTRODUCTION

During the years of disaster studies, several different patterns have been developed to describe stages or events. The historical origin of these cycles is in natural disasters, but they can all be adapted to other types of disasters.

DISASTER CYCLES

The most common approach is to describe disaster response as a series of stages in which the reaction to the disaster ends in the return to normalcy (also conceived to be the situation before the next potential disaster). The various stages can be quite useful heuristic devices to better understand the process. It should be realized, however, that these stages are concomitant and overlapping.

The following is perhaps the most popularly cited series of stages, and is expressed as a cycle. This is the pattern used by the US Federal Emergency Management Agency (FEMA) in its instructional material and courses:

- *Risk assessment.* Basic surveys of a given area are conducted to locate potential disaster hazards.
- *Mitigation.* Contingency plans are drawn up to avoid or to "mitigate" (limit) the effects of a disaster.[1]
- *Response.* As disaster strikes, response agencies are activated.
- *Recovery.* After immediate response is completed, long-term recovery starts. This is integrated with restarting the disaster cycle with renewed risk assessment. (Particularly in countries without an extensive disaster management network, most effort tends to be dedicated to response and recovery, both reactive tasks.)

[1] The British Home Office divides this stage into prevention and preparedness. The latter stage emphasizes planning and exercises.

In this and other cycles it should be remembered that stages are not mutually exclusive. There is overlap.

Despite origins in disasters such as earthquakes, these cycles can definitely be applied to transportation. For example, in an aviation application, *risk assessment* would be possible types of injury in an air crash. *Mitigation* covers such damage-limiting features as fire-resistant upholstery. Following response, *recovery* can include factors such as rebuilding a runway or other crash-caused damage.

Risk is the degree or extent of loss/damage probability after mitigation efforts have been taken against hazards. *Risk assessment* is the result of considering existing risks.

RISK ASSESSMENT

There are three stages to risk assessment (Wang, 2000): (1) identification (identifying what risks are possible), (2) estimation (probability of occurrence and estimation of possible damage), and (3) evaluation (weighing the decision of mitigation efforts should be undertaken).

Many types of damage can be quantified, thus allowing a cost-effectiveness decision regarding mitigation. Damage, of course, is not limited to equipment replacement; it must also take into consideration factors such as *down-time* with no production. That down-time is more readily calculated in a production factory, and less quantifiable in a service-oriented operation. What defies a cost-effective and probability breakdown are the prospects of human injury and loss of life. Although Wang proposed this three-stage analysis regarding offshore incidents, it is applicable to a wide range of disaster response scenarios.

An example of down-time in transportation is the diversion of an aircraft for an emergency landing due to a possible problem. If, in retrospect, the landing was justified, the pilot is lauded as a hero. If, however, subsequent aircraft examination shows the landing to have been unnecessary, ground fees and down-time can be significant. There have been instances in which these costs played negatively concerning the pilot's position in the company.

The assessment of risks and the taking of mitigation decisions are not totally objective procedures. Even when certain risks are identified, people are not always willing to undertake mitigation measures. Sometimes the reason is cost. Sometimes the reason is pure personal convenience. A very simple example is automobile seat belts. It is quite clear that seat belts are an effective mitigation method in car accidents. It has been shown that they reduce both serious injury (50%) and death (45%) (US NHTSA, 1999). Yet, numerous drivers and passengers do not use their seat belts. In one survey conducted by the US National Highway Traffic Safety Administration, 67% of drivers supported regular seat belt use "a lot," but 48% of those same drivers admitted they used seat belts less than half the time (15% rarely/never, 33% only sometimes).

Kletz (1996a) points out that smoking is another known risk that is ignored by a large (but decreasing) percentage of the general population. It is his contention that the public ignores or artificially minimizes risks when they are:

- Self-imposed (such as smoking)
- Under the user's control (driving)[2]
- Natural (earthquake, since it is beyond human control)
- Familiar (side-effects of known drugs)
- Benefit (positive benefits of dangerous chemicals)

[2] Even though there can be fines for certain actions or non-actions.

The same public reacts in disproportion to the threat when confronted with:

- Past negative experience
- Fear
- Breaking of social norms (murder, which results in many fewer deaths than traffic accidents)
- Numbers killed (size of an aviation accident rather than *per person* probabilities)

On the other hand Kletz (1996b) defines an expert's analysis of risk and mitigation as a mathematical determination of probabilities and mitigation costs. Thus, there is a basic contradiction between popular readiness to effect mitigation measures and expert analysis. It is difficult for government to enforce mitigation methods against popular will, even when it is clear that government measures are correct.

Risk assessment is a sustained and ongoing process. Mitigation is, in fact, a work result when risk assessment is clear enough to allow for action. Mitigation steps are, themselves, evaluated on the basis of risk assessment. Disaster impact and response do not bring risk assessment to a halt. Rather, disaster impact creates a new situation. There are new hazards that must be assessed, some of them with pressing immediacy. Building inspection following an earthquake is an example of risk assessment during the response segment of the cycle. Risk assessment really does not follow recovery; it is a guiding principle within recovery. It should accompany the reconstruction process, rather than following it when a new situation has already been constructed.

MORE ABOUT DISASTER CYCLES

The idea of a response cycle is a simplistic model that can be effective as a teaching tool; however it has limited value as an accurate description of reality.

The stages of disaster can be operative concurrently, because those stages are interrelated; they are not independent entities with one stopping and the next following.

Another alternative disaster cycle is:

- *Warning.* Realization that the threat of a disaster is impending. For example, a warning might be a plane preparing for an emergency landing.
- *Impact.* When disaster strikes.
- *Emergency.* The immediate emergency response.
- *Recovery.* Reconstruction and return to normalcy (realizing that the past is never really restored to just the way it was).

Powell and Rayner (1952) suggest another system: warning, threat, impact, inventory, rescue, remedy, and recovery.

The above descriptions of "cycles" are based upon a government perspective, stressing actions as viewed by *government* agencies. It is common that government and official agencies concentrate on their disaster roles to the extent that they overlook how people react, even though it is that reaction that saves most lives in the response/emergency stage.

There are many definitions of "victim." Here the intended meaning is those who suffer direct physical or significant monetary damage from a disaster. A *victim-oriented* system offers:

- *Threat.* The duration of a threat can be constant. There is, for example, a generally tolerated low-level threat, such as ocean travel.
- *Warning.* At this point the threat reaches the stage in which immediate action (e.g. evacuation into lifeboats) is required.
- *Impact.* Victims experience a wide variety of feelings such as helplessness, celebration of having survived. This is often followed by the disaster syndrome estimated to affect some 75% of disaster victims: typical symptoms are being dazed, psychologically numb, apathetic.
- *Psychological relief.* After danger has passed, many people begin to experience behavior such as the desire to talk or act emotionally.
- *Post-incident reaction.* Once the victim has learned how the disaster has affected him (loss of property, family, and friends), he often feels remorse and sadness, sometimes leading to feelings of guilt. This stage can range from several months to a year or two.

There are also various methods of determining the *severity* of a disaster. Each method is a valid estimate from a different perspective:

- *Property damage.* These estimates tend to be very general, since exact values cannot be determined (US FDAA, 1978). Some property damage is never reported through "official" channels, and values can be tailored to insurance claims. Damage to infrastructure and ecology is too often estimated in terms of immediate "repair" and cleanup costs, even though the true costs are often intangible and long-term.

- *Fatalities.* The difficulty with this measurement is that it is focussed on one single issue. In the Gander air crash (12 December 1985) there were 256 fatalities. There were no surviving injured to be brought to hospital, nor was there property damage except to the aircraft and its contents. The numbers of fatalities are also not immediately known. From a treatment perspective, injuries can be more difficult than death due to the cost of rehabilitation, both in monetary and humanitarian terms, that can continue for many years.

RESPONSE SEGMENT OF THE CYCLE

It is often possible to review an incident and determine which official organizations responded to the incident. These responders should be divided into immediate work (fire, police, civil defense, emergency medicine) and secondary work (legal, financial, economic, construction, rehabilitation). Further inquiry can sometimes determine the number of persons and major equipment used. These estimations tend not to include volunteer involvement. Sometimes it is possible to estimate with relative accuracy the cost of the response according to selected parameters.

This book concentrates on the response (government-oriented but not restricted to government) and impact (victim-oriented) stages of disasters relating to transportation. That does not mean, however, that risk assessment and mitigation can be totally ignored. They are part of the overall picture of disaster reality. It is not proper to consider one phase of response out of context with a general description of disaster-related activity.

Risk assessment must take into consideration several different types of disaster, some of which require unique response programs. The typology is, of course, an over-simplification, since many disasters fall into more than one category. To cite just one example, it is not infrequent that an airplane crashes into utility lines causing fires (accidental disaster).

It is overly simplistic to lump together disaster response as a unified plan with universal applicability. A train derailment in China is different from a train derailment in Europe. A train crash in Upstate New York is different from a subway problem in New York City. An airplane crash in Iran is different from an airplane crash in Japan. Although the physical dynamics might be very much the same, local factors can make the response quite different. Those factors can be level of training, experience, terrain/topography, resources available, or religious/cultural requirements.

Some disasters devastate infrastructures, some kill people, and some cause logistical problems. In some cases the nature of the disaster precludes movement of those very supplies needed for disaster response. A crash at a one-runway airport might preclude arrival of a plane carrying needed response supplies. Yet, there *are* certain factors that *do* link the events.

CONCLUSION

It is the purpose of this book to highlight much of the background, methods and problems relevant to transportation disasters that strike in many parts of our world. Stress is on the response phase of the disaster cycle.

Only when the disaster cycle is understood by disaster managers can they begin contingency planning, since that planning must be adapted to phases of the cycle.

CHAPTER 6

PREDICTING DISASTER AND EMERGENCY PLANNING

INTRODUCTION

Even the mere thought of disaster is a threat to our sense of security, and just conjuring up thoughts of death and destruction can be unnerving. This reaction amongst citizenry explains part of the government reluctance to raise the issue of disaster planning. The problem is only multiplied when larger disasters are discussed, and the general populace feels all the more insecure.

In calculated terms, the time and place of disasters are unexpected, but the types of disaster are quite predictable, at least in theoretical terms. Planning is very possible. Some cite the crash of PA 103 in Lockerbie[1] as an "unpredictable" crash (Gilchrist, 1992). Objectively, this was hardly the case, and more of an excuse to justify non-preparation. At the time of the crash some one hundred daily flights passed over the area.

[1] Also site of a train crash on 14 May 1883, leaving 13 dead. The Dumfries and Galloway police constabulary which handled the disaster also dealt with the 1953 *Princess Victoria* accident that left 128 dead. (On 31 January 1953 the car ferry, the *Princess Victoria*, left Stranraer bound for Larne, Northern Ireland and sank.) Even the latter incident was beyond institutional memory by the time of the Pan Am crash.

On 21 December 1988 Pan Am Flight No. 103 en route to New York from London exploded at 31,000 feet. The Boeing 747-121 aircraft broke apart and fell on the Scottish town of Lockerbie (UK) at 19:02:50 hrs. The 259 people aboard the flight and 11 residents of Lockerbie were killed. Victims ranged in age from one month to 79 years, and they came from 21 different countries.

CONSIDERING HAZARDS

For some reason, until recently overhead flights have not been considered possible dangers when planning possible disaster response. Tom Reardon, commander of the Oyster Bay (NY, USA) Volunteer Fire Department, described an air crash in his area as "about as probable as seeing a train derailment in a city without a railroad" (Meade, 1990), despite the numerous area overflights in the New York JFK Airport holding pattern. Then Avianca 052 crashed in that very same area.

On 25 January 1990 Avianca Flight No. 052 from Colombia was in a holding pattern for almost an hour and a half before it was cleared to land at JFK Airport. A report of low fuel in the aircraft was not understood by the control tower. Cause of the crash: the plane ran out of fuel.

Simply, predictability is the outcome of accurate threat analysis, taking mitigation into consideration. Reardon was quite right – there are no train crashes in towns without railroads. His error was in threat or risk assessment. The error was all the more significant in that his area was under an established flight plan. It was not the case of a plane flying over an area not usually accepting overflights.

The disaster at Hillsborough Stadium, however, is a classic example of a known risk covered by a symbolic and inadequate disaster plan and not really foreseen (again a lack of proper planning).

On 15 April 1989, 96 soccer fans died in a crowd crush at a Liverpool vs. Sheffield match in the Hillsborough Stadium in South Yorkshire (UK); 400 supporters suffered injuries. The South Yorkshire Police had directed fans who arrived just before kick-off to enter the already overcrowded stand.[2] Some 10,000 fans tried to enter the Leppings Lane entrance that was equipped with only three gates and seven turnstiles. As the crowd swelled, a crush situation developed.

[2] Overcrowding of sports stadia is a recurrent phenomenon. Overcrowding (25,000 spectators instead of the authorized 15,000) was the reported cause of the Mottaqi Stadium incident in Sari, Iran on 6 May 2001.

According to the Taylor Report investigating the incident, the police officer in charge of the soccer event failed to visit the stadium before approving the plan; and the fire department computer did not specify the stadium location, causing delay in the arrival of forces (Walsh, 1990).

PRACTICAL PLANNING FOR HAZARDS

Applying that conclusion to aviation, computers must be programmed to allow for general addresses. When an aircraft falls in a farmer's field or in a totally uninhabited area, the address of the incident cannot be listed as "123 Main Street." The computer must allow for much more general entries as responders are dispatched. (In the Atlanta Olympics there was a delay in dispatching responders because of computer programming and the inability to enter an exact street address for part of the Olympic Village.)

Another point made by Walsh (1990) is that just because a plan worked "last time," there is no reason to assume that it will work "this time." Sometimes plans must be changed and adapted to such routine factors as day of the week, hour, and weather (the latter bringing to mind that all disaster plans must have flexibility to apply to a changing situation). A basic cause of the Hillsborough disaster was that road construction 80 kilometers away caused delays in the arrival of fans, resulting in crowding. Response to incidents at airports must also take traffic into consideration, both in sending forces *to* the airport and *from* the airport. (Sometimes arriving responders can create traffic problems for emergency vehicles such as ambulances coming from the airport.) Phrased differently, the causes of disaster sometimes can be traced to factors well removed from the disaster site.

Mere existence of a plan is insufficient. Regarding the Loma Prieta earthquake (some 75 kilometers from San Francisco [USA], 17 October 1989), Sgt Curtis Shane of the United States Park Police reported (Shane, 1990), "The emergency plan was nowhere to be found, so we decided to handle the incident [the earthquake] as a major crime scene." At least one air crash was handled on a similar basis. This illustrates that a disaster plan must not only exist; it must be readily accessible, known to users, and drilled at frequent intervals, so that it can be implemented immediately.

One police department surveyed was apprehensive that planning for a catastrophe would cause panic, particularly since its jurisdiction was near an army installation, and the response plan included interface with the military. The entire document was given a security classification (rather than just a classified annex or chapter) and locked away (and unavailable to many of those who needed to be familiar with its content). Emergency plans must be kept readily available.

PARAMETERS OF PLANNING

The *size* on an event for which one plans is a basic issue. In concrete terms, does one plan for a repeat of the Tenerife disaster (the world's most costly aviation disaster in terms of human life)? Or does one plan for a small accident involving only a limited number of fatalities and injured?

On 27 March 1977 in Tenerife, a Pan Am Boeing 747 jetliner collided with a KLM 747 on takeoff. 583 people lost their lives. All 248 aboard the KLM airplane were killed as were 335 of the 396 aboard the Pan Am flight.

Auf der Heide (1989) suggests a medium approach, planning for moderate-size incidents with the ability to expand or retract response – a sort of "best of both worlds" approach, if feasible. Wright (1977) postulates that 120 casualties is the threshold between accident/possible disaster and definite disaster. He was talking from a medical perspective (again an example of situation analysis from a particular point of view). Skertchly and Skertchly (2001) raise some of the problems involved when disasters are truly large.

There is a place in disaster literature for considering certain small incidents that cannot be considered disasters, since they demand full disaster response in their initial response stages. On 2 November 2000, for example, a bomb exploded in the trunk of a car parked in Central Jerusalem, and two people were killed. This can by no means be called a disaster, yet until the true dimensions of the incident were known, there was full disaster deployment by emergency response services. In a similar vein, on 20 June 1997 an area-wide power outage affected Newark (EWR), New Jersey (USA) Airport. This required the implementation of disaster-related programs and provided the opportunity to test the emergency generator system and other response mechanisms, yet the incident certainly was not of disaster proportions.

PROBLEMS OF PLANNING

Most bureaucracies and workers are oriented towards immediate and tangible results. In ordinary government planning, one prepares for a specific program with a known time schedule and a quantifiable ending. Disaster planning is different. It does not have demonstrable results that can be shown to a manager or to the public, if there is (hopefully) no disaster. It is hard to depict prevention activities in the planning stage as the cause of a disaster not occurring.

Many disaster plans are effective, but too many are theoretical statements committed to paper without any practical utility. They might provide formal evidence of fulfilling a bureaucratic requirement that there be a plan, however they can even complicate response to an incident by dictating unrealistic measures.

For example, until the late 1980s one major airline equipped all of its out-of-country stations with a manual that instructed personnel to break into crashed aircraft with an axe (and they provided the axe to do it!). Even the most basic exercise would have shown this to be an instruction out of touch with modern aviation reality. It dates back to the early era of planes crashing *en route*, and search parties combing the countryside looking for wreckage. Plans must be continuously updated, otherwise they have little value. This is an extreme example; however, many paper plans do not take into consideration limits in available space, number of telephone lines, etc. One of the solutions is running emergency exercises, both announced and unannounced, in which plans can be tested.

One airline published a new emergency plan. About six months later the airline's country manager admitted that he still had not read the plan, nor was he really sure of his specific job. The plan certainly will not work if people do not know what it contains.

According to bureaucratic thinking someone must be responsible for all identified tasks (or at least identified key tasks). Identifying tasks should be done in the pre-disaster planning and mitigation stages. Then there can be a logical supposition that in the next step of planning, someone will be placed in charge of tasks. The assignment might not be the best, but at least it provides a contact to whom questions can be addressed.

MITIGATION

Some disasters can be avoided. This can be done through removal of the hazard from a populated area (such as restricting flight patterns over populated areas). In realistic terms, however, not all disasters can be totally prevented. Despite all of our efforts, airplanes will fall. Cars will crash. The key is to mitigate the damage.

GROUND TRANSPORTATION

The police role is significant in dealing with transportation issues, both at the local and regional/national levels. At the local level police mitigation efforts often include route restrictions on vehicles carrying Hazmat.

Another police concern is stopping or mitigating terrorist attacks such as mitigation of potential damage and harm from terrorist attacks using car or bus bombs. One example is the purchase by the Israel Police of mobile computer access devices to identify stolen cars. This device was critical in identifying a stolen vehicle and preventing a car bombing in Jerusalem on 21 March 2001. The vehicle was then exploded in a controlled detonation by bomb disposal experts. There were other mitigation efforts, such as local residents and store owners being told to open windows to lessen the possibility of broken glass. A mechanical remote-controlled robot approached the car so as not to endanger the life of a bomb disposal expert.

Outside the municipal domain, another example of mitigation would be safety belts (cited above) and padded dashboards in cars.[3] These are requirements instituted at the national level.

[3] One wonders how much safety equipment that raises the cost of the car would be included in vehicles if it were not for government requirements.

AVIATION

In aviation, mitigation directives are the responsibility of government oversight agencies, that act based upon the technical findings of previous air accidents and incidents.

Some mitigation efforts concentrate on the design of a particular vehicle. Other efforts relate to the area in which an accident might take place. In aviation, for example, there are strict ICAO (UN International Civil Aviation Organization) rules dictating the availablity of firefighting capabilities with access to an airport runway.

COORDINATION AND PLANNING

In assessing hazards it should be remembered that outside coordination is essential. In an extreme case, the effects of a hazard are not at all restricted to the source area or jurisdiction. The fumes of a Hazmat leak from a railroad car may well come from an area distant from the train itself. Airplanes have crashed into areas not on the flight plan. And, just as hazards do not recognize local jurisdictional boundaries, they also do not recognize international borders. Broad-based coordination is essential.

Economic reality, however, is not simple. There are hazards that are simply not cost-effective to handle at a planning level, primarily due to infrequency and restricted damage. A commercial establishment may make a conscious decision that the theoretical chance of flooding does not warrant a large investment in waterproofing and drainage, since mitigation costs are as high as worst-case damage projections. There are airlines that have unofficially decided not to invest heavily in emergency response programs, because they can be costly. Once a disaster has occurred, the insurance company usually absorbs expenses. On the municipal level, Tel Aviv (Israel) has made the conscious decision not to invest in snow removal equipment, even though it has snowed in the city – the last time being more than fifty years ago!

Sometimes disaster frequency predictions are wrong; in other cases there is a rise in public expectation that demands a new approach, or there is a change in risk. In Jerusalem, for example, for many years snow removal equipment was rare (as were private cars). Only in the 1980s, after newly purchased cars were caught in storms, did the city start investing in snow removal equipment.

If the only damage caused by a hazard is economic, there is room to make a cost accounting before expensive mitigation and response planning is undertaken. When, however, there is a possibility of the loss of human life (or even injury), that frugal approach must be rejected.

Mitigation planning is not merely a paper exercise to show that "everything possible was done." Many mitigation efforts are effective. Two months after the San Francisco area earthquake on 17 October 1989, the Los Angeles International Airport Board of Commissioners initiated an examination to determine the ability of airport facilities and runways to withstand earthquakes; three years later an earthquake did hit the Los Angeles area, and significant damage to the airport was avoided.

PERSONNEL ASSIGNMENTS

One part of emergency planning is the contingency assignment of people to fulfill anticipated tasks. In most situations there is a limited number of people available for emergency assignment. A basic rule is that people should be given emergency assignments as close as possible to their routine jobs.

Emergency assignments should take into consideration not only functions but also skills in daily non-disaster functions. Another factor is psychological profile. People who do not react well under pressure will probably encounter problems in emergency assignment. However, these guidelines are only estimations. There is no absolutely guaranteed method of projecting a person's emergency performance.

There are numerous examples of emergency workers having performance problems in emergency situations. On the other hand there are also cases in which exemplary emergency performance have been stepping-stones to career advancement.

EMERGENCY OPERATIONS CENTER (EOC)

An EOC is both a *function* and a *facility* (Perry, 1995).

EOC FUNCTION

The EOC is the nerve center for a disaster response. It is in this communications hub that representatives of key responding agencies meet to receive field reports, coordinate activities, discuss problems, decide upon action guidelines, and disseminate instructions.[4] Although some of the functions in an EOC are secretarial, other activities require personnel with the authority to make critical decisions.

It is insufficient for only a part of an EOC to be put into operation. Decisions require information, hence the EOC is chartered with the mandate of information collection. If the information is not received and analyzed, then informed decisions cannot be made.[5] If all required participants are not present, there is a flaw in the pre-decision discussion stage. Larger incidents, of course, require

[4] Quarantelli (1979) lists functions of the EOC as: coordination, policy making, operations, information gathering, public information, and hosting visitors. It is agreed that the latter, hosting visitors, is important. One wonders, however, if this is a basic function, or a byproduct of the importance of an EOC.

[5] In the initial stages of disaster there is frequently only partial information available, and rumors abound. It is the role of the EOC to evaluate available data and filter out erroneous information such as rumors.

more manpower; smaller incidents require less manpower. However, all functions must be operating.

A basic exception to this rule is that certain specialized functions are not represented unless they are directly involved. Utility companies, for example, tend to be present only when their service has been disconnected or their facilities damaged.

EOCs exist at all layers of government. It is usually at the municipal level (or airport level in aviation) that activity is most diverse with need for the broadest coordination.

The EOC should not be confused with a Casualty Bureau or Victim Information Center (discussed below). These are two different operations, which can be considered "sub-sets" of the EOC. They use the EOC coordination capability to gain raw data (possible names of victims), then they output data (location and status of confirmed victims) to recipients, including the EOC.[6]

[6] Different operational protocols pass all information through the EOC or allow direct passage between groups with information copies to the EOC.

EOC FACILITY

A well-equipped EOC will have separate working areas for each participating agency. For example, in a municipal model dealing with a Hazmat problem from a derailment, these would include institutional representatives (mayor's delegate, spokesman, personnel from other layers of government, sanitation, public works, housing and welfare, Hazmat experts, police, fire, medical, railroad, etc.), communication lines, and a meeting room. Representatives in the EOC are not restricted to government offices; NGOs (non-government organizations) with significant roles are also present.

Obviously, the exact selection of office representatives to man an EOC depends upon the details of the incident at hand. There are certain routine participants, such as the mayor's office, police, fire, and medical.

Many larger municipalities and government agencies have a ready EOC facility, the existence of which is a visual reminder of disaster preparedness. Even so, it must be remembered that all plans are changeable. Even if there is a fixed site for an EOC, operational considerations can dictate a change of locations. One reason can be the size of the response. The Lockerbie (Scotland, UK) EOC[7] changed venue no fewer than four times, as the response changed phases and mushroomed into a regional response.

[7] The area in which Pan Am 103 crashed was Lockerbie. To be specific, the response was controlled by the Dumfries and Galloway Constabulary that moved its incident operations center to Lockerbie during the initial stages of the response.

Regular offices are not appropriate to serve as an EOC, since one primary objective is to provide a working atmosphere that encourages and facilitates multi-agency interaction and coordination, often on a round-the-clock basis. The use of regular offices also has the disadvantage of encouraging the diversion of attention from the disaster response to routine work (and routine distractions).

In Israel many of the larger municipalities have EOC centers, most often built as part of wartime contingency planning. Although wartime operations are very different, if for no reason other than the extraordinary emergency powers vested in government, the concept of the EOC is quite similar. The physical facility can be easily adapted for civilian disaster use. Its operation, however, requires a different type of thinking.

A different example is Florida (USA), where the state EOC and numerous regional/municipal EOCs are weather-oriented, emphasizing such phenomena as hurricanes, floods and other meteorological events. Here, as well, it is relatively easy to redirect the function of the EOC to non-meteorological disasters as the need presents itself.

Many airports have an "aviation" version of an EOC for air crash response. Perhaps the best known facility in aviation is the British Airways Emergency Procedures Incident Centre (EPIC) located at Heathrow Airport. The services of EPIC are available to subscribing airlines, when they are involved in an air crash relevant to Heathrow Airport.

Many larger airlines have their own corporate EOC that serves the same function of communications hub and coordination point, but only for the internal use of the company. That corporate EOC also keeps track of company manpower needs regarding the disaster.

Even within a municipal government, various agencies can have their own internal EOC, usually operating under a different name to avoid confusion. Perry (1985) refers to these multiple layers of EOCs as *nesting*.

VIRTUAL EOCS

Several Internet EOCs have cropped up in today's age of electronic communications. Although these sites use the term "EOC," they are more public information outlets than true EOCs. A "virtual" EOC is no replacement for a real EOC.

LESSONS LEARNT

In theory at least, disaster planning should be a continuous process, updated periodically as lessons learnt from previous disasters are integrated into revised planning. Learning lessons and applying them are, unfortunately, more often theory than practice. Very few disaster planners make a concerted effort to gather information from incidents happening elsewhere and analyze results as a model for local planning changes. There is also no easy access to this type of information. Only a small minority of planners make the step from passive reading to active learning. Even when they are aware of lessons, they are unwilling or unable to effect significant changes in their own planning. A basic reality is that even when there is a will to effect change, that change takes time to effect.

Taft and Reynolds (1994) point to three levels of learning lessons:

1. organization specific;
2. universal (isomorphic);
3. iconic (superficial knowledge sometimes limited to the mere fact that a disaster took place).

The first stage in the learning process is to obtain a full and accurate picture of what happened. This is often extremely difficult, if not impossible, due to the chaotic nature of the disaster itself and the selective memory of many of the psychologically traumatized participants.

One common mistake is to think that when a disaster experience is analyzed properly, changes or improvements in procedures and equipment constitute lessons learnt. These are reasoned consequences after an overall analysis. They are practical steps *derived* from lessons learnt. For example, a lesson is not that headbands with flashlights should be worn. This is one of the options available from a lesson that rescue workers should have both hands free, even when close lighting is required.

Nor should the search for lessons double as an undercover operation to seek blame for failure and to dole out punishment. If that latter approach is taken, any attempt to drawn constructive conclusions will be counter-productive and will preclude candid participation in inquiry responses.

ValuJet Airlines Flight No. 592 (DC-9-32) en route to Atlanta, Georgia (USA) crashed in the Florida Everglades near Miami, on 11 May 1996. All five crewmembers and 105 passengers aboard were killed.

Lessons learnt also should not be self-serving. Following the 11 May 1996 crash of ValuJet, the US National Transportation Safety Board (NTSB) issued a report stating in part, "Had the FAA [Federal Aviation Authority] required fire/smoke detection and/or fire extinguisher systems in class D cargo compartments, as the [National Transportation] Safety Board recommended in 1988, ValuJet flight 592 would probably not have crashed." Even if this statement is totally correct, it gives the initial impression of a self-serving finding. It also is written as if to say that the FAA totally ignored NTSB findings.

In fact, after initial NTSB warnings FAA dealt with the size and materials of class D cargo compartments. Unfortunately, only on 14 November 1996 (months after the crash) did the FAA decide, "that it would go forward with rulemaking to require fire detection and suppression systems in the cargo compartments of all passenger aircraft" (FAA APA 110-97). If a proper lesson is to be learnt, it is not that the FAA did not act at all on NTSB findings (lessons learnt), but that it reacted only partially and too slowly. Ways must be found to make the system work faster. Even after the ValuJet crash, FAA realized that a three-year phase-in period was needed, "to require retrofit of Class D compartments with fire detection and suppression systems on nearly 3000 older commercial aircraft."[8]

[8] At an estimated $50,000 to $70,000 per aircraft.

When viewed from a more general perspective, it must be realized that government agencies usually work notoriously slowly, often to the consternation of citizens needing or wanting service. It would be unusual if a high-cost program, such as that recommended by NTSB, were to be implemented rapidly without thorough examination. Some of the delay is part of formal procedures. After the recommended change was published in the Federal Register, according to regulations the FAA gave a 90-day period for comments before the change took effect.

After many disasters, people involved in them write articles describing what was done and what lessons were learnt. All too often these articles are personally enhancing and non-critical, if not total fiction bearing little resemblance to the incident in question. In some cases "what should have been done" is not clearly separated from "what was done." These articles are often simply a retelling of events as recalled from personal memory, rather than a critical analysis of response procedures gathered from systematic interviews. That retelling is done from the participant's perspective; rarely does he have a general overview of the entire incident; rarely does he put his own actions into context.

Auf der Heide (1989) points out the reality that all too often lessons learnt are incident-specific. Very rarely are conclusions drawn after reviewing a series of disasters. This overall analysis is more common in academia and much less frequent in government circles, even in regional or national governments where there is a broader base of experience than at the local level. He also points out that even when a personal disaster report is accurate, one does not know whether the situation reported is common to numerous disasters, or whether unique only to the disaster under consideration. Again, this goes back to the lack of multi-incident overall perspective.

This conclusion can be taken one step further. Lessons learnt by whom? Since incident reviews tend to be unit-specific, a lesson learnt by a fire department might not be communicated outwards, even though there can be direct application to a police department. Particularly where a negative lesson is "what we did was wrong", the material is often not transmitted to other localities or jurisdictions (which then are liable to repeat the same mistake). Nor is the communications mechanism necessarily well-refined enough to allow transmittal even when the material is based upon a positive experience – "what we did was right."

SYSTEMS ANALYSIS

What is needed is not a one-agency disaster plan or a series of one-agency plans, but a comprehensive plan that takes into consideration and coordinates all aspects of response. This means planning from an overall perspective, realizing that all responders must act in coordination, as though they belong to the same agency.

There is no one single cause of a disaster; there are numerous causes. In modern society we have built-in multiple safeguards to prevent disasters and mitigate their effects. Disasters are, in effect, systems failures. They are a breakdown of our safety infrastructure. Airplanes have backup systems wherever possible. So do trains and ships. An accident is usually caused by more than one failure. In the same sense, the entire "system" must respond.

Perhaps this is a crucial difference between developed and developing countries. In a developed country the governing "safety culture" is more and more attuned to disaster prevention.

On 2 August 1973 a fire broke out, leaving 50 dead at the Summerland Leisure Centre on the Isle of Man. In the control center the on-duty personnel did not call the fire department. The direction of moving stairs on the escalator was not reversed to allow people to escape. Several emergency exits were locked. Parents added to the confusion by searching for their children (a very natural reaction). In short, the entire *system failed.*

Nor was this the first fire at the Leisure Centre. In 1940 a fire broke out at the same facility, and again the safety system failed. Electricity was turned off, but the generators were never switched on. Doors did not meet fire regulations. Turnstile direction was not reversed to allow exit and the fire alarm malfunctioned (Turner, 1992). Again, the *system failed.*

Although the transportation industry tends to be more safety-conscious than the above example, it also suffers from system breakdowns, particularly in its interface with other disaster-responding agencies.

TQM

Management fads change from time to time. At the time of writing this, Total Quality Management (TQM) is the concept guiding industrial operations and it has made inroads into bureaucratic government operations as well. Its basic precepts were written on business models. Even in government situations models were non-emergency situations. A large part of TQM can be adapted to disasters, particularly in pre-incident planning, however certain readjustments must be made.

If the philosophy inherent in TQM is to be successfully imported into disaster preparation and response thinking, basic precepts must be translated into the operational milieu of disaster and the *disaster subculture*. Such terms as *cost* and *competitive edge* must be redefined for the disaster manager. In many response situations the key is, "Do it!" and cost becomes secondary. *Service* and *customer satisfaction* must be built around a monopolistic situation in which key operators are mandated by law. There must also be a specific listing of what services are to be given, who the customers are, what are realistic expectations, and what are plausible measurements. How does one measure the effectiveness of a fire response when an earthquake has caused damage to access roads, and water mains have lost pressure?

In the private sector competition plays a major role. In a market economy, competition is the threat to existence and the spur to excellence. In government this is not the case (although inter-agency rivalry is certainly a known phenomenon). The roles of police and fire are, for example, monopolistic. In a disaster response one cannot "pick" which police or fire department to call. Even in overlapping geographic jurisdiction (such as areas that have city, county, and state police), responsibilities are well defined. Nor does disaster response to a real incident allow for comparative bidding; that is a procedure restricted to planning and exercises. Urgency dictates the use of available resources, and the only control on their use is theoretical in planning stages.

Applying quality assurance to disaster response is not simple. To date there are no standards defining a "good," "mediocre," or "poor" response, except the unwritten and subjective evaluations of seemingly endless people – each one ready to voice his opinion. It is very difficult to measure a response without basic guidelines that have earned common acceptance.

Disaster management cannot operate according to quality assurance principles unless the component parts of the response also use that management philosophy. After all, most of disaster management is the coordination of different organizations and activities. If TQM (or any other quality assurance program) is to be used, it must be part of *everyone's* basic planning.

In England an opposite tactic has been taken to quality management. Rather than redefining terms to meet the needs of government offices, there have been efforts to transform certain government services into a quasi-commercial management model. This has been implemented to a lesser degree in Australia.

No one is opposed to providing a quality product or service. The first step is setting that as a goal, then defining terms. Only when the basic goals are set can programs be measured in terms of fulfilling objectives. Otherwise, setting work and achievement scales is an exercise in creating movement, but not necessarily activity that will generate better service. In victim identification, identifying "x" number of bodies in "y" time is an example of movement; this must be accompanied by work guidelines that will set procedures to preclude error (i.e. to ensure quality). Following those procedures also explains why it might take longer to record *post mortem* information in one case than in another (e.g. the condition of the mouth area might mean more effort is needed to take the required number of dental photographs). In some cases mere movement might have the opposite effect, sacrificing quality so that "more" can be recorded, or criticizing an employee for what he has not done, rather than recognizing and rewarding him for quality work that he has done.

TQM begins early in the planning stage, where it is common to begin with lectures. Too many instructors talk *about* TQM. They never get past that introduction to talk about *how* to achieve quality. The goal of a good lecturer is to transform passive listeners into enthusiastic activists, and in disaster planning the challenge is all the greater, since the general subject is known to generate complacency, if not apathy.

After all, TQM starts with the worker. It is a work ethic, and not merely a required directive. It is imperative for a lecturer to motivate his audience, then he can pass out material to read. If a TQM program starts by distributing written material, it will be lost by the time most people are convinced enough to read it seriously. This training is particularly difficult in disaster response, since at some stage it must move from being agency-specific to inter-agency.

Starting a TQM program means starting with the worker. If management is interested only in the end product, eventually there will be no workers to produce that final item. Workers, however, are not soloists. An ideal situation is where workers join together as a team with each person adding his contribution. One can veritably say that the aggregate of a team effort is greater than the sum total of its parts. Teamwork also creates the work atmosphere, which is a cardinal factor in determining the work product. This spirit of teamwork is particularly important in a disaster response, where camaraderie amongst co-workers can be a major factor against post-traumatic stress disorder (PTSD).

Man is a complex creation. Motivation is a strong element in quality assurance. A worker must also have a sense of satisfaction, knowing that he has helped others. Merely setting up production-line style tasks and requiring quality implementation will not succeed. Only when the worker is motivated can total quality be addressed. Otherwise quality assurance efforts will be symbolic, allowing a so-called manager to place a check on his work sheet, "Yes, we covered that issue as well."

ASSESSMENT

Another cardinal principle of quality assurance is to conduct a critical review of what was done, so that real lessons are learnt. All too many times malfunction is ignored out of supposed politeness or it is excused "under the circumstances." Politeness is not *what* is raised; it is *how* the matter is expressed, and *why* the issue is discussed (constructive discussion and certainly not vendetta).

CONCLUSION

As shown, it is essential to develop disaster plans; however those plans must be realistic. They must be updated; and they must take into consideration as many lessons learnt as possible. A solid knowledge of hazards is a prerequisite to effective disaster contingency planning. If hazards are not identified in the planning stage, they certainly cannot be mitigated, and response becomes all the more difficult.

So far government planning has been stressed. Private companies, such as transportation companies, must have their own internal plans as well.

CHAPTER 7

PRIVATE SECTOR

INTRODUCTION

Business concerns often meet pre-disaster planning with apathetic reactions, thinking that to improvise at the time of need can be a viable substitute for pre-incident planning. At best they take disaster planning as a cost-effectiveness statistic, gambling that the money devoted to planning and the probabilities involved are overridden by projected profits. This is as true with some airlines, bus, train and boat operators as with certain manufacturers, since these transportation companies act as businesses in every respect (even when they are government companies).

Government agencies tend to follow similar budgetary thinking, unless they have regulatory responsibilities that force a more active role. (At the same time that they take an interest in the disaster response programs of the business that they regulate, they tend to ignore their own internal response and recovery plans.)

One major difference is that in the final analysis there is a certain transparency about government: it is a relatively open operation subject to citizen complaint. There can be pressure exerted on government to prepare properly. That pressure barely exists in the private sector.

PREPARING, NOT PREPARING

This avoidance of disaster thinking is little different from the phenomenon of psychological *avoidance* after an incident, and it is buttressed by a logic sequence based upon:

- it will not happen;
- if it does happen, it will not happen here;
- if it does happen here, it will not be on my shift (not my responsibility);
- in any event we will get through it;
- because we (they) got through it last time.

One of the obvious roles of government is to ensure "disaster thinking" in the operation of transportation companies. For example, a ferry boat company must be licensed and its ferries inspected periodically. Licensing should mean not only ship inspection, but also coordination with emergency response authorities. Proper ship inspection includes review of fire extinguishing equipment, emergency vests, lifeboats, *and* emergency procedures. Those procedures are not limited to the operation of safety equipment; they also include evacuation and rescue. In turn, the company's internal plans to handle fire must be coordinated with local fire, police and emergency medical resources. Disaster thinking means that the private company understands what must be done, even without government regulation.

When new airports or seaports are built, their planning should take disaster scenarios into consideration. Access roads, for example, should be built with use in mind as possible entrance routes for emergency vehicles as well as possible evacuation roads. This means the roads must be of sufficient width and they should be protected from parked cars. All locations within the port should be accessible from at least two directions.

"Disaster thinking" must also be flexible. After all, it is *thinking* and not merely a civilianized substitute for military command and control. Rigid adherence to plans is bound to incur failure. One must adapt plans to developing conditions.[1]

[1] Ledger (2000) postulated that a significant part of the cause of the Swissair crash in 1998 was due to the pilot's rigid adherence to work rules.

This might be fine for planned development for new facilities created from blueprints approved by committees, but it offers no consolation for older facilities built long ago. The problem is particularly acute in Hazmat incidents of trucks entering narrow city streets or barrios constructed by urban squalor and sprawl. Principles of historic preservation now take disasters into account, and tourist revenues can often generate budgets to "improve" what can well be called urban museums. This urban redevelopment can give a partial answer to quick evacuation of victims after a traffic accident, or a disaster of even greater proportion. Outlying slums, however, that are far from the view of politicians and tourists, often receive much less attention . . . until disaster strikes.

INSURANCE

An often overlooked aspect of disaster planning is interface with the insurance industry. Insurance companies must deal with such issues as repair versus replacement, and repair methods. This ranges from a car damaged in a traffic accident to a building hit by debris in an aviation disaster . . . and the aircraft itself, including its occupants.

Since insurance companies have to pay compensation for loss, they are generally involved in aspects of mitigation. Ironically, however, insurers are satisfied that there be a certain low level of damage and claims. After all, only because there are instances of damage do people take out insurance policies. If there were no threat, there would be no demand for insurance.

Aviation disasters are particularly costly, and a spate of accidents can be difficult for insuring companies. That is one of the reasons that aviation insurers seek company protection in consortiums. For example, the La Reunion Aerienne consortium comprises CGNU (UK), Abeille (UK), Generali (Italy), Groupama–GAN (France), and Mutuelles du Mans Assurance (France).

Insurance companies handling compensation claims very often ask questions about licensing and zoning; if a building or business operated against local requirements, this can often be a basis to deny insurance payments. This is again complicated when disaster damage is partially covered by government grants, thus making the government and the insurance industry "partners" in disaster reconstruction.

Another joint insurance–government concern is fraudulent or criminally inflated claims of damage to be reimbursed, claims based upon previously existent damage, or claims that are paid but repairs not done. The National Insurance Crime Bureau (NICB) has estimated that "property/casualty-based insurance fraud costs Americans $20 billion annually. Barry Zalma, a California (USA) attorney and insurance fraud expert, claims that insurance fraud 'takes as much as $100 billion, or more, every year from the insurance industry in the United States.' It is hard to establish a more exact figure, since most fraud goes undiscovered and undetected. In contrast Hurricane Andrew [the costliest insurance event in US history] resulted in $17 billion of damage." (NICB Report, 1997). The Canadian Coalition against Insurance Fraud estimates that fraudulent claims represent between 10 and 15 per cent of claims paid out, making it the second leading source of criminal profits in North America – second only to illegal drug sales.

The number of claims in disasters can be overwhelming, and it is relatively easy to hide fraudulent claims amidst the sheer number of cases. Natural disasters, of course, are more prevalent, wider in scope, and generate more claims than transportation disasters. According to Wood (2000), the ice storm in Eastern Canada in 1998 resulted in almost a million individual insurance claims.

Sometimes the actions of individuals cause suspicions of insurance fraud. According to the Guardian (27 May 1998), a passenger on the ill-fated TWA 800 flight "took out life insurance policies worth several million Swiss francs in the weeks before the plane crashed in July 1996." (For the record, examination of his luggage and other investigations showed no sign of carrying explosives or criminal intent.) There were numerous reports that the pilot who committed suicide by crashing a SilkAir flight also bought insurance with a payment value of $3 million.

On 19 December 1997 SilkAir Flight No. 185 (Boeing 737-300) crashed into the Musi River near Palembang, Indonesia. According to an *Aviation Week and Space Technology* report, the deaths of pilot and 103 other people aboard the flight were caused by the pilot's suicide.

It is important to record the names of all persons involved in an incident, whether they report being injured or not. As the US Fire Administration (1996) reports, "Public transportation systems have had many false insurance claims from people who were later shown not to have been on the conveyance at the time of the mishap." This applies to virtually all types of transportation.

There are misunderstandings about the role of insurance in disaster response. Having an insurance policy does not protect against damage. That policy can give *financial* compensation to the insured, but it gives little else. Lost luggage following a disaster can be paid for by an insurance company (according to the details of the policy, which many people do not read until they try to collect). That does not necessarily "replace" the content. Life insurance policies often pay a double premium for violent death (as in a disaster); however no sum of money can ever really compensate for loss of life.

Unrecovered property presents a particular problem for insurance companies. After the Swissair crash, claims for unrecovered property totaled a full kilogram of diamonds, 4.8 kilograms of jewelry, and about $1 million in cash ([Toronto] *Globe and Mail*, 23 December 1998, p. A4). To guard against fraudulent claims many insurance policies do not cover cash and set an upper limit on the value of any one single item, unless it is specifically covered in a special policy rider.

CONCLUSION

The private sector operates upon commercial considerations; in that environment disaster planning is all too often placed on a "back burner." Part of the calculation in planning or not planning is the role of insurance in offsetting the cost of losses. A mistake can also be made in thinking that disaster planning is limited to response. Even if the probability of a disaster is extremely low, just one event can be devastating. Private companies, therefore, have a real interest in their own efficient disaster response, and in the response of government agencies.

CHAPTER 8

WHEN DISASTER STRIKES

INTRODUCTION

Advance warning precedes some natural disasters such as hurricanes or tsunamis. Transportation disasters are different. If warning precedes a transportation disaster, it is usually only minutes before a potential incident. An example would be a radio message from an aircraft declaring an emergency. Preparedness, therefore, must be continuous with response forces always on call.

PREPAREDNESS

Despite the need for constant preparedness, this is not always possible. That preparedness is sometimes "on call" and not physically present. Sometimes key personnel are not available for one reason or other, and can be reached only by telephone (if at all). Therefore, back-ups should always be part of disaster planning.

HISTORICAL CASES

This disaster does not relate to transportation, but it is a part of American history, and a clear lesson can be learnt.

What has come down in history as the "Great Chicago Fire" happened on 8 October 1871. (Contrary to popular credence, the legendary "Mrs O'Leary's cow" was probably innocent.)[1] Firemen had put out a major fire the previous day, and as was their custom, they celebrated at the bar. When the fire broke out, many of the firemen reportedly were still suffering the after-effects of the "celebration."

In the middle of the night prior to the late afternoon Air Florida/Metrorail incidents there had been a four-alarm fire, thus exhausting the responding Washington, DC fire department. It has been contended that this tiredness hindered the response by reducing alertness and increasing reaction times.

On 8 September 1934 the *Moro Castle*,[2] a passenger ship carrying 558 passengers on a trip from Cuba to New York, caught fire off the shore of Spring Lake,

[1] History records that Mrs O'Leary hastily moved out of Chicago shortly after the fire.

[2] Named after the historic Moro Castle in Cuba.

New Jersey (USA) (the rescue operation was directed from Asbury Park) and 134 people died in the fire. Captain Robert R. Willmott could not take command of his ship. The night before he had died of a heart attack in his stateroom.

It is not only personnel that need back-up. The need for reserve equipment can also be a problem in disaster response. For example, the Loma Prieta earthquake was a natural disaster. In this case, transportation played a key role in the response. Following the earthquake the Highway Patrol helicopter nearest the Cypress Freeway (collapsed area with large number of casualties) was already operating with the low fuel warning light showing. At the same time the Oakland Police's only Enstrom was out of commission and being repaired (Michaels, 1990).

With all of this said, it must be remembered that many emergency services work around the clock. They handle more than one incident at the same time. When Washoe County (Nevada, USA) Police sent out the first message regarding the crash of Galaxy Flight 203 (21 January 1985 at 0104), the police were on skeleton-force night manpower. The nearest potential unit for response was in the process of handling an armed robbery. As another example, The MGM Hotel in Las Vegas (USA) caught fire on 21 November 1980. The first fire call came in at 0717. The three top managers of the hotel were busy on another project – a fishing trip off the California coast. The police were also busy on another more serious project – at 0600 they had run an operation yielding 33 drug arrests.

There is a direct parallel in transportation. Not only at MGM did managers take a vacation at a time that proved inopportune. British Airways set up the Emergency Procedures Incident Centre (EPIC) to handle communications after an air disaster. When Pan Am 103 crashed in Lockerbie, the EPIC director was on vacation in Central America (John Nichols, personal conversation). On 6 July 1989 a terrorist forced a Tel Aviv–Jerusalem (Israel) bus off the road, killing 13; at the mortuary a responding fingerprint technician waiting for bodies to arrive was called away and asked to deal with a murder that had occurred elsewhere in the Tel Aviv Police District.

Improvement in a system is always desired, however service cannot be compromised in the change-over to new equipment. The Kansas City (USA) Skywalk collapse happened just when the local EMS system was converting to the Public Utility EMS Management Model; some services had merged into the new system, while others were still tied to the older system.

HAZMAT INCIDENT

The following Sahuarita, Arizona (USA) incident serves as an example not only of an off-hour transportation disaster, but also of an incident happening in a very small jurisdiction. Local police and fire services retained command, but they were totally overwhelmed by several hundred personnel from numerous agencies (personal correspondence).

A train derailed in Sahuarita, Arizona (USA) on Tuesday, 2 January 2001 at about 2000. A police force established in 1997 and staffed by eight commissioned officers (including a chief and deputy as supervisors, a detective and a K-9 handler) covers the area. On the night of the derailment there were two officers on duty with about six years' experience between them. Sixteen to nineteen rail cars fell off the tracks; five or six of the cars contained sulfuric acid (99 per cent pure) and several held red phosphorus residue. Fortunately, only three of the sulfuric tanks ruptured; about 5–10,000 gallons were released.

Curiously, the incident commander took responsibility and determined a 1.5 mile area to be voluntarily evacuated, based on prevailing weather and computer projections of possible Hazmat-caused contamination. This evacuation was against the judgement of railroad representatives, who did not think even voluntary vacating of the area was necessary.

HAZMAT GUIDANCE

The first step in dealing with a Hazmat incident is to identify the hazardous material present. This can be done initially from shipping manifest or through rapid-reaction Hazmat testing kits (Kaplan and Almog, 1983). Kit results should always be verified by full laboratory testing (Almog and Glattstein, 1997), particularly if legal proceedings are being contemplated.

Basic reference material can be found in the North American Emergency Response Guidebook (NAERG) and the USFA Hazardous Materials Guide for First Responders.

Notice of evacuation was given at 2100 and delivered by some 20 policemen knocking on doors. Notification through the media was judged to be inappropriate and impractical given community demographic characteristics. Notes were left when residents were not at home. About ten people chose to remain; 96 were evacuated to a high school cafeteria where the American Red Cross had food and hot drinks; about 65 went to relatives' homes in the area.

Numerous lessons can be learnt from these examples, but what stands out in all of them is that disaster can happen at any time, particularly when that time is the "least convenient" for responders.

NOTIFICATION OF DISASTER

The news that there has been a disaster is not always received from expected sources. The Nassau County Police Department (NY, USA), for example, first heard of the Avianca crash from a resident who called the emergency switchboard to report a plane down next to her house. This "rather unusual" call came before official notice of a missing aircraft from air traffic control at JFK Airport.

During this author's police career, the first news of at least two disasters was received from radio bulletins and not through established police channels. Rather than criticizing the relative slowness of official communications, it is more constructive to plan deployment taking into consideration the fact that many responders will hear of incidents through the media before notification is made in official channels.

FIRST ON SCENE

The immediate role of the first professional on site is always to *survey* quickly what has happened, then notify appropriate management. Even a superficial survey enables responding agencies to judge manpower and equipment requirements and dispatch responders accordingly. In addition to avoiding unnecessary traffic problems en route to the disaster, calculated dispatch can keep reserve personnel fresh for later shift work. If all workers are summoned for immediate service, assigning manpower to subsequent work shifts becomes difficult. These considerations are part of *resource management* (auf der Heide, 1989), a technique used to maximize upon response capabilities, while at the same time avoiding *over-response.*

The problems inherent in over-response relate to more than manpower usage and shifts. The presence of unneeded responding agents can add both congestion and confusion to a disaster site. This has the further consequence of preventing an accurate assessment of the response situation and real needs. Over-response also depletes resources for another emergency that might occur concurrently or soon thereafter.

"UNIQUE-PROBLEM" VICTIMS

Some victims can be unusually anxious about their situations. Matters are not always as simple as they might initially be expected.

For example, on 27 June 1976 an Air France A300 flying from Tel Aviv to Paris was hijacked and landed at Entebbe, Uganda. Passengers and crew were rescued in a commando raid. Then one of the hijacked victims had to explain to his true wife why he was traveling with another "Mrs" – his secretary.

Northwest Flight 255 that crashed upon takeoff on 16 August 1987 in Detroit, Michigan (USA) (Warnick, 1987) had a similar problem of passengers posing as married couples. The flight was also overloaded both with more crew and more passengers aboard than was permitted (Silverstein, 1992), causing the airline problems of explanation.

Sometimes the problems involved fall into the criminal domain. On 26 May 1991 a Lauda Airlines Flight Number NG 004 (Boeing 767-300ER) en route from Hong Kong to Vienna, Austria with a stopover in Bangkok crashed in Ban Dan Chang District, Suphan Buri Province, Thailand. Ten crew members and 213 passengers were killed. One passenger was traveling on a bogus passport.

On Avianca Flight 052 that crashed on approach to New York Kennedy Airport there were two narcotics "mules" (passengers smuggling concealed cocaine); both were arrested in hospital after examination (Maher, 1990).

These are just some examples to show that when dealing with people, one must always expect the unexpected.

ELECTED OFFICIALS

No matter what the job position of elected officials, they feel the need to be seen, heard, and given information. The higher their position, the more visibility (direct and implied) they want. After some disasters, particularly natural disasters where no *direct* blame can immediately be levied,[3] elected officials can serve as a rallying point for public morale.

Visits to airport crashes tend to be "non-issues," since the disaster scenes are often far from public view. Following terrorist incidents, however, elected officials can find themselves the focal point of public anger about the perceived breakdown of security. This is even more pronounced when the incident is one in a series of terrorist attacks (although the same visits, if made later after tempers have cooled, can give a boost to public morale).

These visits by public officials should be carefully planned and not made as a spontaneous reaction to a disaster. Very high profile officials who are recognized by the public usually need some type of protection, at least after terrorist incidents. This is particularly problematic for the police, since they must protect the VIPs as well as handle the disaster.

One purpose of these VIP visits is to gain a first-hand appreciation of what is happening, in order not to run the incident from a detached and distant location. The escorting entourage, however, can have the effect of distancing

[3] Few disasters, even natural, are not eventually blameable on someone who could have prepared better or who was actually negligent.

the official from the event. Discussions with on-site responders also tend to be formal exchanges with no critical content. Hence, if the official cannot overcome these difficulties and justify his visit as a learning experience, it is best for him to postpone his visit. He will, in fact, learn more from videotapes and well-written analyses of response successes and failures.

POST-INCIDENT REPORTS

It is important to keep log books and similar records during a disaster response. The time to start is at the very beginning of a disaster. These records make the inevitable post-incident reporting and analysis all the easier.

Very often extraneous considerations influence disaster reports and statistics. Reducing damage to "protect" a positive image of the locality is very common. In one known case local government officials tried to reduce disaster statistics with the intention of maintaining the image of safety. A train operated by Great Northern Railroad was covered by an avalanche in the mountains of Washington State (USA) near Wellington on 1 March 1910. The company tried to keep the death toll to 96, so that the number of fatalities would not qualify the incident as the biggest rail disaster to date in US history. The figures were juggled and the official number (96) was released. As fate would have it, another body was recovered after snow thaw (making it 97). The disaster then became the biggest in US railroad history (Moody, 1998).

Although government reports do include statistics, these reports tend to be dry publications issued long after the incident (and after the general public loses interest). The most influential reports are not the carefully written documents that are the fruit of long months of work. They tend to be those reports cited by the press – in the confusion of the incident, by reporters who are unable to evaluate the material properly because it is outside the scope of their professional expertise and experience.

There is a basic conflict in incident reporting by government authorities. It is quite natural that a local municipality would want to estimate disaster severity – the key is estimate, since absolute objectivity is extremely difficult if not practically impossible – in the most minimal terms possible. Many newspapers, however, prefer sensationalism inasmuch as that is what sells copies. Thus, there is apt to be a difference between official and media reports.

ECOLOGY

In modern society it is axiomatic that care be given to disturb our surroundings as little as possible: this means an effort not to pollute the atmosphere, and to leave an area as clean as possible after using it. A disaster response, therefore, is not over merely when the last ambulance or fire truck leaves. That response is over when the area has been cleaned as completely as possible and restored to its regular use. Ecology is not only care for animal and plant life. It is also concern for people, their property, and their privacy.

Following the 1987 Amtrak crash the usually quiet local neighborhood in Chase, Maryland (USA) was transformed into the center of national interest.

On 4 January 1987 three Conrail freight locomotives traveling between Baltimore and Harrisburg at above the permitted speed entered the path of an Amtrak passenger train near Chase, Maryland (USA). The crash killed 16 people and injured 170.

No fewer than 1276 press interviews were conducted with residents, and four community meetings were held to deal with the situation. Two years later some residents were still complaining of psychological after-effects and of ecological damage to the neighborhood. In some ways the response caused more damage than the accident (litter, lawn damage, use of homes, use of private telephones – costs hard to recover with no clear agency or group responsible to compensate).

Disasters do cause damage. Litter and rubble at the scene are relatively easy to clean, although the job can be time consuming and costly. It is much more difficult to deal with the ecological ramifications of a disaster. The crashes of TWA, Egyptair and Swissair off the North American coast all involved aircraft early in flight with large amounts of fuel that spilled in the ocean which had to be cleaned up. Other disasters have caused fires that destroyed wooded areas.

On 17 July 1996, TWA Flight Number 800, a Boeing 747 aircraft en route from New York to Paris, exploded off Long Island, New York shortly after takeoff. All 230 people aboard were killed. The cause of the crash is still a matter of contention.

On 31 October 1999 Egyptair Flight Number 990 crashed off the Massachusetts (USA) coast after having taken off from New York bound for Cairo. The 217 people aboard, including 30 Egyptian military officers, were all killed when the Boeing 767 crashed into the sea.

On 2 September 1998 at 2231 Swissair Flight Number 111 (MD-11 manufactured by McDonnell Douglas) from New York to Geneva crashed off Peggy's Cove, Nova Scotia (Canada) killing the 215 passengers and 14 crew members aboard the flight. The pilot dumped fuel in the Atlantic in preparation for an emergency landing (Ledger, 2000).

CONCLUSION

The expected must be expected at a disaster scene. The unexpected must also be expected. This is part of preparedness. Then responders are in the best position to handle a disaster, using both their own forces and auxiliary assistance from other organizations.

CHAPTER 9

USING THE MILITARY

INTRODUCTION

Using military forces can be a tempting solution to a manpower shortage problem in a disaster. Using soldiers, however, is not without its problems. Sometimes it can be a very problematic solution.

MILITARY CAPABILITIES

There are many different circumlocutions for the word "army." Sometimes the soldier-image is camouflaged by calling it National Guard or Home Front. Despite linguistic acrobatics, one must realize that the military does not have a civilian orientation. Their role in civilian operations can, therefore, be problematic. Under *certain* circumstances, however, the military can play a positive role in a disaster response, particularly when the incident is large scale or when special logistics are required.

In a disaster the biggest strength of an army is generally in logistical support – the ability to mobilize a large number of vehicles (land, sea, and air) and to deliver materials. In one case in Israel (12 December 2000), for example, a bus tumbled off the road down steep hills and into a valley. The army provided night-time lighting and ropes to assist responders to climb down the incline. Conversely, the weakness of military support is that armies tend to be cumbersome, requiring extensive lead-time unless specialized units, always on call, are utilized.

Another military strength is information gathering (i.e. intelligence). During a disaster, updated information about the current situation is often required. Some of that information comes from military sources such as aircraft (Perry *et al.*, 1993) and has to be integrated with information received from civilian sources. The military, however, has to make the mental switch from enemy targets to civilian needs. Another information-gathering application is the search for ships at sea or airplane survivors in naval or non-populated areas.

POTENTIAL PROBLEMS

A standing army in a non-conflict situation is heavily administrative, with either symbolic operational manpower or the minimum number of soldiers required to hold defensive positions in case of attack until reservists can be drafted and sent to the front. Thus, the reality is that, particularly following a sudden disaster, an army is at its lowest strength when assistance is most needed – in the immediate two or three hours after a disaster. In times of disaster, using the army is further complicated by the fact that soldiers can, themselves, be victims or missing persons.

It is often forgotten that most military supplies are inappropriate for the civilian population.

- Although an army might be able to supply a large number of blankets and tents to disaster victims, storerooms do not have children's clothes nor medicines for serious diseases that would preclude army service.
- In many cases armies have been used to provide food in the first stages of a disaster response; army food is oriented towards MREs ("meals ready to eat"), but it has limitations in that it does not include special diet foods (e.g. salt or sugar free).
- Career military doctors do treat the wounded and can administer first aid, but they do not routinely handle medical problems associated with the young or elderly, nor with the handicapped.
- There are also legal questions involving protection against suit when a military doctor renders treatment (a) to a civilian, or (b) for a problem outside his true field of expertise.

Thus, if military logistics are used, this must be in support of a civilian-generated assistance program and not a military program initiated for civilians.

CIVILIAN VS MILITARY CONTROL

It is a mistake to put the army in control of a civilian situation, unless the basic infrastructure has been totally destroyed and lawlessness can be thwarted only by the imposition of martial law.[1] Even despite this not being the case, martial law has been prematurely declared in numerous incidents. Its effect can be more negative than positive.

[1] This is presented as a theoretical option that is to be found only in the most extreme cases, if at all. A possible example might be a massive earthquake generating extensive secondary problems. This does not apply to transportation disaster. Yet there have been transportation incidents, particularly at sea, where military forces have assumed command.

EXPECTATIONS FROM THE MILITARY

An infrastructure breakdown is hardly the case in a transportation accident, but even so, there have been cases of bringing in the military. At Lockerbie, for example, there was no breakdown of civilian infrastructure. Rather, civilian manpower was severely over-burdened, thus raising the issue of military deployment.

Armies are meant to work independently using only their own resources. That is a key element in battle strategy where outside assistance is impossible, and it is a general underlying principle in military thinking. Soldiers do not have a clue about the other players and their functions in a disaster situation, nor are they trained to work with "outsiders" (non-military personnel). On the other hand, civilian agencies can ask for outside help or not, depending upon circumstances. Since civilian agencies routinely work with coordination (even to the extent that the cooperation is *so* routine that it is taken for granted), they do not necessarily see their agency as the dominant organization that must take control (although it is common to see post-incident reports claiming that the writer's agency "did all the work").

The very method of thinking in an army is different from a civilian situation. As Avrech and Dluhy (1997) point out, a military commander knows that in war the first casualty is peace-time planning. An innovative approach is often what is required. In the civilian sector, however, pre-incident disaster plans are expected to set the parameters for response. Even if innovation is in order, the bureaucracy expects compliance and will instinctively reject deviation.

It is also a mistake to utilize individual soldiers, except in special situations. Armies are not individual soldiers; they are designed to be viable units, and they are taught to function as units. These units must be mobilized, equipped and positioned to do what they are trained to do. If artillery units are taken to rescue people from the wreckage of a crashed aircraft or from damaged buildings, they will be performing a highly technical task without training.

It is similarly a mistake to think that the army is needed for command and control. That is a military solution with poor application to the civilian sector. Even in smaller disasters one should realize that there is a danger of speaking in civilian terms, but thinking in military terms. Some people *speak* of civilian "emergency *management*," yet at the same time they really *think* in terms of military "*command and control.*"

CONCLUSION

Calling in the military to assist in a large civilian disaster should not be a natural reflex. It should be a calculated decision made after weighing both pros and cons. One of the basic issues to be considered is whether the military has received proper training to handle the civilian tasks involved.

CHAPTER 10

TRAINING

INTRODUCTION

Reason dictates that all plans must be taught if they are going to succeed. In other words, training is a critical aspect of all disaster planning. The stages of training are: identification of disaster roles, training needs, persons to be trained, and methodology. No plan is so simple that it does not need to be taught or at least presented and explained before implementation.

ROLE

In a disaster responders are given a role – an accumulation of tasks. In the most technical sense a switchboard operator answers incoming telephone calls and forwards them to the telephone extensions most appropriate to give answers. Answering calls and forwarding them are two technical tasks. The "role" of that switchboard operator is much broader.

One can take, for example, the case of a telephone operator employed by an airline. In a disaster situation the operator must be aware of the employer's operations and decide if an incoming query is at all within the purview of the company, or if the call should be diverted elsewhere. If the answer is positive, the operator must record basic information or direct the call to an appropriate extension, depending upon apparent urgency and importance. Training must be both at the level of technical tasks *and* role requirements. If a broad definition of need is not taken in designing training, people will wind up being able to function only in a very limited manner.

NEEDS

It is axiomatic that there must be a "needs assessment" before training modules can be developed. Going back to the switchboard operator, training needs might range from:

- instruction in telephone manners (not to criticize current functioning, but to take into concern the ability to deal with callers under stress and sometimes overly sensitive in interpersonal exchanges)

  to:

- re-entering the line of a person on hold to see if he wants to wait or leave a message.

These needs are not disaster specific. They are general needs in playing the role of switchboard operator. That is exactly the point. In designing training even the *non-disaster specific* must be taken into consideration, so that the entire response effort can function properly. In the above example it well might be that the operator in question does not deal with people in stress or with an overloaded switchboard during routine work.

There are certain subjects, such as occupational *safety* in a disaster area, which should be taught to all persons potentially involved in a disaster plan. It is generally preferable to separate managers from line workers in these training sessions.

PERSONS TO BE TRAINED

It is not always clear who should receive training, or which courses they should attend. Sometimes those who need to be trained do not have key disaster response tasks. Rather, they are interaction (Wilson, 2000) or support personnel.

METHODOLOGY

There is no "one way" to train. A basic rule of teaching is to expose students to learning through as many *senses* as possible. People should *hear* material (through lectures). They should *see* it through reading, films or demonstrations. They should *feel* it through hands-on exercises. A training program should take all of these possibilities into consideration as students are taught. Varying the method also has the additional effect of preventing boredom. Training methodology has its own literature.

TESTING

In many work scenarios it is not considered "polite" to test students in a training program, particularly when the students are more senior than the instructors.

Testing, however, should not be compared with a child's bringing home a report card for signature. It is simply verification to the instructor that the material has been learnt. The grade certainly need not be part of the employee's personnel file. (The fact that he finished the training successfully *should* be a matter of record so that organizational training status reports can be compiled.)

Testing can be formal (written), informal (oral), or practical (performance of a task) according to the specific situation.

LESSON PLANS

In a professional training module the subject and teaching goals are defined in a written lesson plan. After the lesson plan is drawn up, a more detailed course outline is then written, showing the points to be taught in their order. The lesson plan should also include:

- required reading;
- suggested reading;
- visual aids to be used;
- field trips.

Regarding trips, taking people to particular places important to the training can be an effective didactic tool. If, for example, the instructor wants to explain response deployment stations at an airport, one approach is to show them on a map. Another approach is to visit each site and explain who will use each area; this is particularly effective with those not totally familiar with the area.

Methodology is often dependent upon the number of people in a class. Smaller groups allow for more flexibility in methods used.

When a full lesson plan is prepared, the instructor is in a better position to know when he has covered all material. It also makes it easier later on for other instructors to give the same course.

ATTENDANCE

Sometimes employees feel that training, particularly for contingency planning, is taking them away from their "real" job. This can degenerate into signing an attendance sheet, then disappearing under the guise of answering a supposedly urgent telephone call.

A good way to start a training module is to counter this attitude by explaining the importance of the training. An off-site venue (not in the building where those to be trained are employed) can also help, in that it is not so easy for students to wander off. A sign-up sheet distributed during the training and not upon entry in the morning is another method of dealing with the issue.

EVALUATING INSTRUCTORS

Another common taboo is to evaluate the presentations of instructors, but there could be nothing so mistaken. If an instructor is to be entrusted with teaching, there must be some gauge of effectiveness. It is not sufficient that an instructor know the material; that goes without saying. He must also be able to deliver it effectively. Student evaluations, always with the caveat that their names *never* be released to the instructor, can be an important tool. It is a method for instructor self-evaluation. More important, such evaluation sheets provide clear indication about re-inviting speakers in the future.

CONCLUSION

Training is an important part of disaster preparation. The training program must be well planned, taking into consideration both general disaster subjects and those specific to the jobs people will be doing.

CHAPTER 11

WARNING AND EVACUATION

INTRODUCTION

Warning and evacuation are real possibilities in certain transportation disasters. Understanding and handling these issues should be part of training programs.

PLANNING

Gathering information needed for mitigation and response must start long before the onset of a disaster. As Granger and Johnson (1994) have suggested, there must be a disaster "information culture." There must be a routine flow of information, so that people are disaster-aware and mitigation-aware. People must understand what happens in a disaster and how to react. The net result: if a disaster culture is prevalent, citizens are more likely to obey warnings, instructions, and evacuation orders when they are announced.

"Citizens" is a broad term that is most often used to denote residents of an area. It can also be applied to company workers, who have to make the mental "switch" from routine jobs to emergency assignments. That switch in thinking is another by-product of a disaster culture.

Evacuation can be considered a step in mitigation. By removing people from an area, disaster managers are eliminating the possibility of physical damage.

THREATS

There are several different types of threats that can result in evacuation: meteorological (flood, storm), geo-physical (earthquake, volcano), conflict and hostility, industrial accident (Bhopal, India; Chernobyl, Ukraine), and transportation. Concerning the latter, the most common reasons for evacuation are:

- Hazmat
- fire
- structural damage

The most common reason for evacuation in a transportation disaster scenario is the leaking or potential leaking of Hazmat cargo (from either train, truck, or crashed aircraft) near populated areas. Some of these phenomena are preceded by warning (e.g. *possible* Hazmat leak); others strike suddenly without notice (e.g. *actual* Hazmat leak). If people are sensitized to the dangers and implications of these disasters, it is easier to evacuate them during a transportation-related Hazmat incident.

CASE HISTORIES

Mississauga, Ontario (Canada). At 0004 on 10 November 1979, 24 cars on the 2km-long Canadian Pacific Train No. 54 derailed after one car loaded with toluene overheated, causing the wheels to lock and the axle to melt. A fire ensued, then spread to cars carrying propane, causing a large explosion. Other cars carrying caustic soda, styrene and toluene were damaged and leaking. Police then evacuated a large part of southern Mississauga.

Crescent City, Illinois (USA). On 21 June 1971, 15 cars of a freight train derailed. Nine of those cars carried liquid petroleum, and others carried propane gas. There was an immediate evacuation because local inhabitants understood the threat. Although 90 per cent of the neighboring business district was destroyed, there were no fatalities (Richardson, 1996).[1]

[1] It is estimated that 70 per cent of the population in the United States lives within 5 miles (8 kilometers) of a Hazmat site or a transportation route for Hazmat.

Propane gas in rail cars was also involved in an evacuation in Flint, Michigan (USA) on 21 January 2000. Six hundred people were taken from their houses, two schools were closed, and Interstate Highway 475 was shut down. The latter meant extensive re-routing of traffic.

Evacuation should not be resisted in times of necessity. When people do not realize the risks involved and remain in an endangered area, the price in human life can be dear.

Kingman, Arizona (USA). In 1973 fire started as workers were pumping LPG (liquid petroleum gas) tankers. Twelve fire fighters lost their lives. The threat was not properly understood, and as a result the crowd that gathered was not held far enough back. In all 95 spectators were injured (Richardson, 1996).

Warnings must be given in very clear language, since people can react emotionally without critical evaluation. That necessity for precise and appropriate language is necessary in all types of disasters.

In a non-transportation example, in August 1955 there was heavy water pressure on a Delaware River dam due to the heavy rains of Hurricane Diane. The Port Jervis, New York (USA) Chief of Police announced that letting out water would lessen pressure on the dam. This was misunderstood as a warning that the dam might break, and many local residents fled the area without due cause.

TO EVACUATE OR NOT?

When the decision is made to issue a disaster warning, the public goes through four stages of reaction:

1. collection,
2. evaluation,
3. decision,
4. implementation (Williams, 1964).

That is to say, in the first stage the intended recipient *collects* or receives the disaster warning message.[2] He then *evaluates* the message, generally in terms of perceived personal relevance. Evaluation often includes consultation, usually with family, close friends, and interested parties, such as an employer. This consultation can entail extensive use of telephone and cellular phones (thus tying up lines needed by emergency responders). This evaluation, however, is not necessarily totally rational. It is done in an emotion-charged atmosphere, and even if information collection is methodical, that does not mean that data analysis and evaluation is also objective.

[2] This is in the best case. Very often the disaster warning is not received by all due to a range of problems such as hearing problem, sleep, television/radio off, etc. In some cases the message is received. It is just not understood, at least as the sender intended.

It has been estimated that potential victims with extensive education tend to take warnings more seriously than uneducated persons. This is a problematic supposition. During the Iran–Iraq War the "better educated" Iraqis distrusted local news services and listened to the BBC and Voice of America, which did not carry local air raid warnings. Thus, these people did not hear the emergency instructions (that they distrusted anyway, since they were most often a false alarm).

The next stage is to *decide* what, if anything, should be done. This decision can include rejection (fear of looting, dismissal of seriousness, desire to remain to help, search for family members). Finally, the message recipient *implements* whatever decision he has reached.

Evacuation, therefore, is a person's calculated decision, even if emotional and psychological considerations outweigh objective thought. It is very rarely a panic reaction.[3] Movies depicting crowds fleeing in panic from impending disasters, whether from fictitious monsters[4] or more plausibly from an invading army, are creations of Hollywood and not depictions of any reality. Yet, these same movies (even when showing crowds fleeing Godzilla or other unrealistic creatures) sustain the myth of flight in panic.

[3] Panic for this purpose is defined as emotional fright and those actions resultant therefrom. Panic reactions are typically emotional, if not irrational, and not based upon careful consideration.

[4] E.g. the classic King Kong (1933) or the more modern Jurassic Park (1993 and sequels).

In reality it is rare to *panic* during an evacuation. This has been shown extensively in numerous studies, from the evacuation of buildings to the evacuation of ships (Okerby, 2001). In a sense one can say that sometimes public officials "panic." They have been known to delay evacuation notices unnecessarily out of fear that they will engender panic amongst recipients.

There are several other factors that contribute to the decision concerning evaluation of the warning and the decision to act upon it. Source of information can be a deciding factor. These sources can be divided into formal, informal and media. An example of a formal source would be a government agency such as police, fire or the municipality. Informal sources are family, friends, and neighbors. In general, the more personal the source of warning (e.g. family, friends), the more serious the warning is perceived to be.

The warning is also perceived to be more serious when direct signs of the disaster (such as burning buildings not far off) are first seen. Such signs are not always seen immediately in a Hazmat incident, particularly if the hazardous material has not yet started to affect the area.

When the danger is imminent and obvious, then the person involved goes through the same mental process as described, but this time in a matter of seconds.

Frequently people are motivated by personal advantage. They will make a calculated decision (often simultaneously based upon rational and emotional criteria) concerning what they perceive to be in their best interests. It is a mistake to counter emotional reactions only with rational arguments.

Sometimes that "imminent and obvious" nature of the danger is a misperception.

PSYCHOLOGICAL REACTION

If warning is over a long period of time, such as the slowly rising waters of a river, there is a chance that people will *normalize* their situation, in effect disregarding the warnings. This explains the attitude of residents of the neighborhood at the end of the Guatemala City Airport runway; they have "normalized" their situation by psychologically ignoring or even dismissing risks of overflying aircraft. Given this rationalization, it then becomes difficult, if not impossible, to implement any serious pre-incident disaster mitigation program. (It often *is* possible to mitigate further damage once an incident has already occurred.)

In some incidents *normalization* can be considered a variation of *denial*, as victims of an incident try to convince themselves that danger indicators are related to non-threatening causes. Shaw (2001) cites an example of airplane cabin smoke being dismissed as routine cigarette smoking.

WHERE TO RELOCATE

In the best case evacuation is transfer to the residence of a close friend or family member outside the danger zone. In such cases the guests are welcome in the short term, but not in the long term (a period of time the definition of which is influenced by the type of disaster, length of the perceived threat, family situation, and other intangible and real perceptions). In many instances there are evacuation centers set up by governmental authorities. In these cases, even when the stay is limited to several hours, it is important to register evacuees, so that they can be traced by relatives (or by responders needing information).[5] Although luxurious hotels are seemingly a pleasant option, experience has shown that evacuees are not in the frame of mind to enjoy the full facilities of the hotel, as they worry about their home and personal situation. These hotels also tend to be cut off from sources of information other than the general mass media. Thus, although the hotel theoretically offers comfortable surroundings for the evacuation, in practice the experience is less positive than would be at first thought. (Conversely, "the grass is always greener on the other side of the fence." Those evacuated to centers or lower quality hotels often feel that other facilities would have been better.)

[5] American Red Cross Form 5972, "Disaster Shelter Registration," has been designed for this purpose. In cases of evacuation it is also desirable to have available telephone numbers of evacuees (and a sufficient number of telephone lines to handle their incoming and outgoing telephone needs).

People must be told very clearly where they should go. In Bhopal some people sought shelter in the streets. This might be appropriate for an earthquake, but certainly not for a Hazmat incident.

Even when someone does decide to evacuate, it is not always possible. Roads may be blocked, transportation may not be available, or the situation may have deteriorated to the extent that it is safer to remain put.

It is not sufficient to issue a formal "proclamation" ordering action such as

Bhopal, India stands out as one of the most costly Hazmat disasters of the 20th century in terms of its toll on human life. On 3 December 1984 methyl isocyanate (CH_3NCO), used in the manufacture of pesticides, leaked from the manufacturing plant of Union Carbide in the city.[6] A chemical reaction caused expansion and a general release of poisonous gas. Although many of the workers in the plant were able to put on protective masks, people in the area were not so fortunate. They were not properly included in the emergency plan. Almost 4000 people died.

Even inside the plant there were problems with the safety manual. It was written in English. The language of the workers was Hindi.

Although the problem was centered in an industrial area, transportation was also involved. As Silverstein (1992) reports, at least one train was waved through the town without stopping. Other trains were quickly re-routed.

[6] The International Chemical Safety Card suggests that after a spill the affected area be evacuated and ventilated. Complete protective clothing must be worn. It is a point of conjecture to what degree that protective clothing was available.

evacuation. Unless there are tangible signs of imminent danger, another aspect of the decision-making process is related to the "marketing" of instructions. Thought should be given to how instructions are presented. In general terms, the wording of an evacuation instruction should be chosen carefully to provide information, yet to prevent fear; to show a threat, but to affirm that the government is in control. Thought should be given to the most appropriate person to deliver the message – a neutral expert (e.g. police), a known official (e.g. mayor), or an impartial professional (e.g. broadcaster). In the Israeli experience, instructions during the Gulf War were given by the army spokesman, who became a known and trusted personality on local television screens.

It is very difficult to evacuate a person without his family, unless he is certain that they are safe, or the danger is so imminent and obvious that there is no real choice. This is particularly true in case of parents/children. Pets play a smaller (yet still significant) role in deciding whether or not to evacuate.

Methods used to notify the population that evacuation is required or recommended very much depend upon local circumstances. Radio and television broadcasts are to be taken for granted. Loudspeakers in vehicles running through a neighborhood might not wake those asleep, particularly on the upper floors of high-rise buildings. Many weather situations leave residential windows closed, again complicating notification. Knocking on doors and ringing bells is time consuming; a non-answer also does not guarantee that no one is home. Notifying the handicapped can also be problematic.[7] In short, a variety of methods must be used to reach a maximum number of people.

[7] One estimate has suggested that as much as 40–50 per cent of the US adult population has some type of disability.

The problem of conveying emergency messages is not only related to evacuation. The same problem exists when issuing other emergency instructions, such as warnings to turn off utilities or to stop using water.

CROWD DYNAMICS DURING EVACUATION

There has been research on the evacuation of stadia. Instead of applying them only to stadia, one should apply the conclusions to other public facilities that contain large numbers of people who must be moved very quickly. That can happen in airports and other transportation terminals, just as it can happen in stadia. For example, an explosion in Frankfurt Airport in the 1980s caused a large evacuation. On 30 October 2000 Boston's Logan International Airport was evacuated (including passengers aboard aircraft) when guns were found at a security check (Los Angeles Times Internet Edition, 30 October 2000). On 18 January 2001 part of the Schiphol Airport (Netherlands) terminal was evacuated after there was an explosion in a bathroom and an ensuing fire. There have been, of course, other evacuations, particularly due to terrorist threats.

When a facility is certified for maximum capacity, it is not only seats and waiting space that are important. A maximum evacuation time must be set, then an estimate must be made as to the number of people who can be evacuated in the specified time given exit ramps and gates. Fruin (1981) measures crowd mobility and its ability to function by density.

Space per person (sq. ft)	Behavioral result
25	Normal walking speed; passing
10	Much slower
5	Maximum safe density; shuffling; restricted movement
3	Brush up against each other
2	Dangerous; psychological pressure

STAYING HOME

There have been incidents in which people have refused mandatory evacuation orders. This is, perhaps, the most difficult situation for disaster managers to handle.

In recommended (not mandatory) evacuations the situation is much easier to handle. Making a conscious decision to remain home even after an advisory evacuation warning is not without some benefit. These people who remain often serve as an unofficial information resource for those who have evacuated. They are able to report specific information to those who have evacuated (such as condition of property), and their reports tend to be trusted by listeners.

Research shows that those who evacuate are more susceptible to psychological problems than those who stay home.

RETURNING HOME

Returning *during* an incident can be more dangerous that declining a non-compulsory evacuation.

Returning home *after* a mandatory evacuation or after a Hazmat incident should be permitted only after certification that the danger has passed.

When there has been the possibility of basic damage to a building (such as a vehicle or airplane striking the structure), even short-term re-entry should be made only after inspection.

CONCLUSION

Warning and evacuation instructions are not simple steps. They are complex issues involving proper wording, timing and delivery. There are associated questions such as mandatory vs recommended evacuation, and what to do with people refusing compulsory orders.

When there is an evacuation, records should be kept of those leaving. Thus, when disaster does strike, subsequent search and rescue operations in the area are simpler. There is no need to search for people who have relocated.

CHAPTER 12

SEARCH AND RESCUE

INTRODUCTION

There are basic differences between Search and Rescue (SAR) following an earthquake or fire, and SAR following a plane or ship disaster:[1]

- In the former, the site is known and SAR focuses on rescuing victims.
- In the latter, the site is often not known when operations begin.
- In the latter, an approximation of the number of victims can be deduced, when there is a passenger list available.

[1] Searching for an airplane is relatively easier than looking for a ship. Yet, in searching for the Lauda aircraft which crashed in northern Thailand, rescue teams first located (and initially misidentified) a different aircraft that had crashed earlier.

TRANSPORTATION DISASTERS

Rail disasters are relatively easy to locate, since they are limited to areas adjacent to tracks; however experience has shown that rail rescue may need unique equipment not used in road accidents (the usual focal point of fire and police departments) and not necessarily readily available (US Fire Administration, 1994).

Perhaps an extreme example of SAR to locate a missing aircraft is the case of Juliet Delta 321 that crashed in the Antarctic on 4 December 1971. Only 16 years later was the plane, buried in the snow, salvaged at a site 1400 kilometers from McMurdo Station, Antarctica.

In all cases SAR efforts cannot be measured as a cost-effective operation. The classic question is how much should be invested to save one human life. It is ironic that much more is invested *per capita* in *post facto* Search and Rescue than in pre-incident mitigation.

INTERNATIONAL EFFORTS

The United Nations Department of Humanitarian Affairs maintains its International Search and Rescue Advisory Group (INSARAG) in Geneva, Switzerland. Their main functions are to act as a clearing-house for SAR efforts, establish standard procedures, and disseminate lessons learnt. Their work, however, is not at all oriented toward transportation incidents, nor toward naval search in the case of sunken ships.

METHOD

A basic work guideline of SAR operations is to locate spaces (e.g. empty space under furniture, walls not totally collapsed) in which people can survive. A person caught under the total collapse of floors above him will be crushed, with virtually no chance of survival. One technique of finding survivors is by the use of Trapped Person Locator[2] (TPL) equipment, designed to detect noise possibly caused by a person. This will not work when looking for fatalities.

[2] Example manufacturer is Elpam Electronics Ltd, 28 Hacharoshet St, Or Yehuda 60375, Israel. (Mention does not indicate endorsement or recommendation.)

Another technique is the use of dogs for searching, as was done in the LAPA (Boeing 737) crash upon takeoff from an airport near Buenos Aires (Argentina) on 31 August 1999; at least 71 of the 103 persons aboard the plane were killed.

Another method of SAR is to remove all debris from an area, eliminating all possibility that survivors still remain. This approach is appropriate to routine operations in small accidents, but in a disaster it usually means spending excess time on marginal possibilities, rather than dedicating limited resources to saving as many people as possible. Once it has been determined that life saving is no longer possible, the "total clear" method can be used to ascertain that all bodies have been found. This generally does not apply to transportation accidents, unless buildings are also involved.

RELATIVES

In some SAR operations it is recommended that security personnel accompany rescuers to protect both personnel and equipment. Survivors have an understandable interest in reordering priorities so that their families will be the first objectives of SAR operations. This can, unfortunately, express itself in trying to prevent SAR or other response teams from proceeding further to work at other locations. In transportation accidents the sites are quite defined, and it is easi*er* (but certainly not "easy") to secure the area.

Experience has shown that families will remain at a disaster site for hours if not days, waiting for the missing to be recovered from the rubble. In societies where property is not generally insured, retrieval of belongings can also be an issue of central concern to survivors. No matter how one tries to maintain distance, the responding SAR teams are always under the scrutiny of the survivors. This can mean implied pressure to work even beyond reasonable shifts. For foreign teams to explain procedures, translators are often needed. The families are often an important source of "intelligence" information, identifying buildings totally destroyed and providing maps of rooms with suggestions as to where residents are to be found.

SAR OPERATIONS

There is considerable disagreement as to the optimum time for a SAR or other disaster response work shift. Generally, shifts are planned to last 12 hours; however this is not necessarily realistic, given the need to overlap at both ends of the work period, and the lack of any relaxation atmosphere (often leading to renewed work outside an official shift).

ABOARD AN AIRCRAFT

People do evacuate airplanes when instructed to do such by the aircraft crew. There are, however, incidents in which the crew did not instruct evacuation even in the face of imminent danger, either due to their incapacitation or to their misunderstanding of the situation.

After a plane crashes with ambulatory survivors it is frequent that crew or even passengers take a decision-making leadership role in assisting others to leave the aircraft, but there is no steadfast rule that this will be the case. There are documented examples of passengers: (1) evacuating prematurely, (2) opening a door that let flames into the cabin, (3) remaining seated awaiting instruction, and (4) saving themselves by evacuating. There is no one correct response that covers all situations.

The phenomenon of emergent leadership is known in many other types of disasters. The main difference here is that the leadership can often be very limited in time (from the time of the crash until the arrival of professional forces – if on field, then only a matter of minutes).

There are numerous stories about people weighing a decision to evacuate a ship such as an ocean liner. There the size of a large vessel affords a certain amount of anonymity, and as long as a passenger holds out for himself some hope of survival if he stays aboard, evacuation prior to the arrival of rescue forces is not necessarily undertaken. Descending into a small lifeboat is, in itself, a danger. Buses, given their smaller size, afford no such anonymity; leaving the bus poses no decision-making problem.

BODY REMOVAL

It is not uncommon that it takes a significant amount of time to extricate bodies, particularly after building collapse or following air crash when multiple sites are involved. This is usually thought of in terms of disasters involving buildings (e.g.

earthquake, explosion, and structural collapse), and there are certainly enough examples. This was the case following both the AMIA explosion[3] in Buenos Aires, Argentina and the *Loma Prieta* earthquake. In both cases extrication of the last body came more than ten days after the incident. Nor is the extrication effort necessarily over after the last body is recovered. It can also take additional effort to ascertain that no more bodies are to be found or that remaining bodies are to be left.

[3] Although commonly called an "explosion," this was, in fact, an "implosion." The bombing caused a suction reaction that caused building floors to rise, turn over, then fall. The result was an inversion in the order of certain floors with lower floors resting above higher floors.

Extrication, however, can be a difficult problem in transportation accidents. This is complicated by the need to use non-sparking extrication tools when fuel is present and constitutes a fire threat.

Maritime incidents are the most awkward in terms of extrication. Often it is necessary to work through the inner sections of a ship, sometimes underwater, to reach victims. However, on-the-ground incidents can also be difficult. The Lauda crash occurred in difficult jungle terrain. Thai International crashed into a hard-to-access mountain slope. The Avianca 727 that crashed in Cucuta, Colombia on 17 March 1988 caused a landslide that buried a good part of the wreckage.

Fires also can require a delay in extrication activities. Following the 1949 fire on the *Noronic* in Toronto (Ontario, Canada) Harbor, only on the following day did the ship cool down sufficiently to allow extrication; 111 bodies were found (107 with no personal recognition possible). In total, search of the burnt vessel took two weeks (T. Brown, 1952).

In a disaster there often are bodies that require extensive extrication efforts. This means that not all bodies are recovered at the same time. The process of body recovery can take several days, if not longer.

In fact, people can die in various phases and areas of a disaster. In the *Noronic* fire, for example, in addition to those deceased recovered from the ship, two persons died in hospital, one person was dead-on-arrival, and five drowned in the harbor.

Basic theory is that all of the bodies should all be collected together in the incident mortuary, thus allowing the medical officer in charge to have a total picture, as much as possible, of the incident. This, however, is not always possible, since sometimes the external pressures to release identified bodies does not allow the waiting time until all bodies are recovered. There are, of course, cases in which all bodies are *not* recovered.

At the disaster site, once a body has been removed from wreckage, it can be necessary to carry the cadaver on a stretcher to the collection point. Gurneys might theoretically require less effort, however they do not provide a practical solution, since their wheels can be caught in debris and in mud (which even on a clear day is often created by water from the extinguishing of fires).

In many cases body parts are to be found in rubble or amongst ashes. In these instances it is preferable to have a forensic anthropologist or pathologist accompany the extrication effort to help identify body pieces. Hinkes (1989) reports that at the Gander, Newfoundland (Canada) crash, remains from different bodies were initially mingled in the same bags, a problem that had to be solved later by anthropologists.

CONCLUSION

Although Search and Rescue are most often thought of in disasters such as earthquake, these procedures are also used in transportation disasters – both in terms of vehicles and the buildings into which they crash.

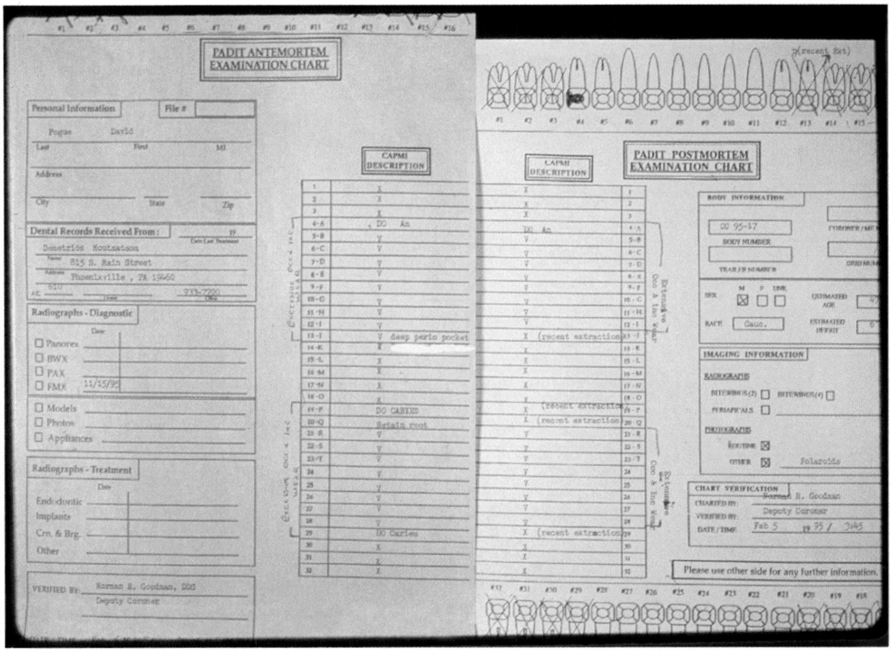

PADIT ANTEMORTEM EXAMINATION CHART

Personal Information | File #
Pogue David
Last First MI
Address
City State Zip

Dental Records Received From: 19
Demetrios Koutsatsos
815 S. Main Street
Phoenixville, PA 19460
610 933-7200

Radiographs - Diagnostic
Date
☐ Panorex
☐ BWX
☐ PAX
☐ FMX 11/15/95

☐ Models
☐ Photos
☐ Appliances

Radiographs - Treatment
Date
Endodontic
Implants
Crn. & Brg.
Other

VERIFIED BY: Norman E. Goodman, DDS
Deputy Coroner

PADIT POSTMORTEM EXAMINATION CHART

BODY INFORMATION
CC 95-17
BODY NUMBER
SEX M ☒ F ☐ UNK ☐
RACE Cauc.
ESTIMATED AGE
ESTIMATED HEIGHT

IMAGING INFORMATION
RADIOGRAPHS
BITEWINGS (2) ☐ BITEWINGS (4) ☐
PERIAPICALS ☐
PHOTOGRAPHS
ROUTINE ☒
OTHER ☒ Polaroids

CHART VERIFICATION
CHARTED BY: Norman E. Goodman
VERIFIED BY: Deputy Coroner
DATE/TIME: Feb 5 1995 / 3:45

Please use other side for any further information.

Photo 1

Odontogram designed by Dr Norman Goodman of Philadelphia, Pennsylvania.

(Photograph: N. Goodman)

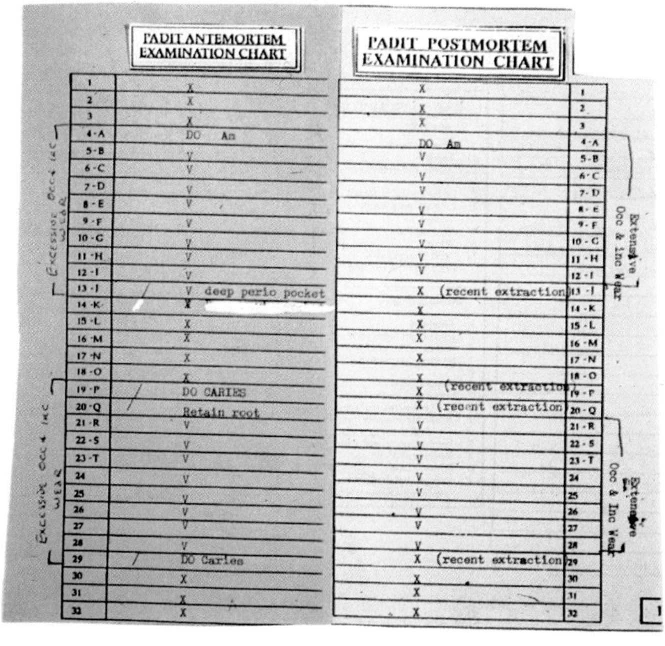

PADIT ANTEMORTEM EXAMINATION CHART		PADIT POSTMORTEM EXAMINATION CHART	
1	X	X	1
2	X	X	2
3	X	X	3
4-A	DO Am	DO Am	4-A
5-B	V	V	5-B
6-C	V	V	6-C
7-D	V	V	7-D
8-E	V	V	8-E
9-F	V	V	9-F
10-G	V	V	10-G
11-H	V	V	11-H
12-I	V	V	12-I
13-J	V deep perio pocket	X (recent extraction	13-J
14-K	X	X	14-K
15-L	X	X	15-L
16-M	X	X	16-M
17-N	X	X	17-N
18-O	X	X	18-O
19-P	DO CARIES	X (recent extractio	19-P
20-Q	Retain root	X (recent extraction	20-Q
21-R	V	V	21-R
22-S	V	V	22-S
23-T	V	V	23-T
24	V	V	24
25	V	V	25
26	V	V	26
27	V	V	27
28	V	V	28
29	DO Caries	X (recent extraction	29
30	X	X	30
31	X	X	31
32	X	X	32

Excessive Occ + Inc Wear (teeth 4–13; 19–29, antemortem)

Extensive Occ & Inc Wear (teeth 4–13; 21–29, postmortem)

Photo 2

The odontogram is designed so that ante mortem and post mortem data can be placed side by side for comparison.

(Photograph: N. Goodman)

Photo 3

Freeway collapse following the Loma Prieta earthquake.

(Photograph: courtesy of David Fowler)

Photo 4

Car caught in the collapsed freeway.

(Photograph: courtesy of David Fowler)

Photo 5

Crash of USAir Flight No. 427.
Evidence collection is an extensive operation.

(Photograph: N. Goodman)

Photo 6

Documentation found at the USAir crash.
The documents were not found in the pockets of a victim.

(Photograph: N. Goodman)

Photo 7

Crash of Thai International in Nepal. Clothing was placed on lines to dry and for families of survivors to identify.

(Photograph: J. Levinson)

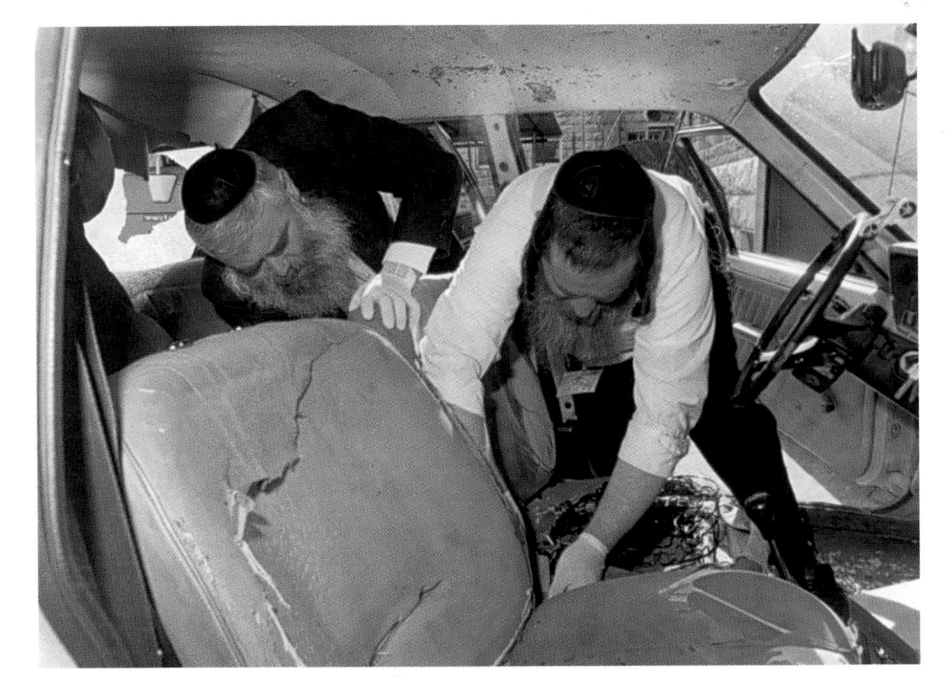

Photo 8

Some religions require burial of blood. Volunteers collecting blood of a deceased after a fatal road incident.

(Photograph: Israel Police)

Photo 9

Bus tumbled into a valley. Responders had to pull themselves back up to the road using ropes provided by the Army.

(Photograph: Israel Police)

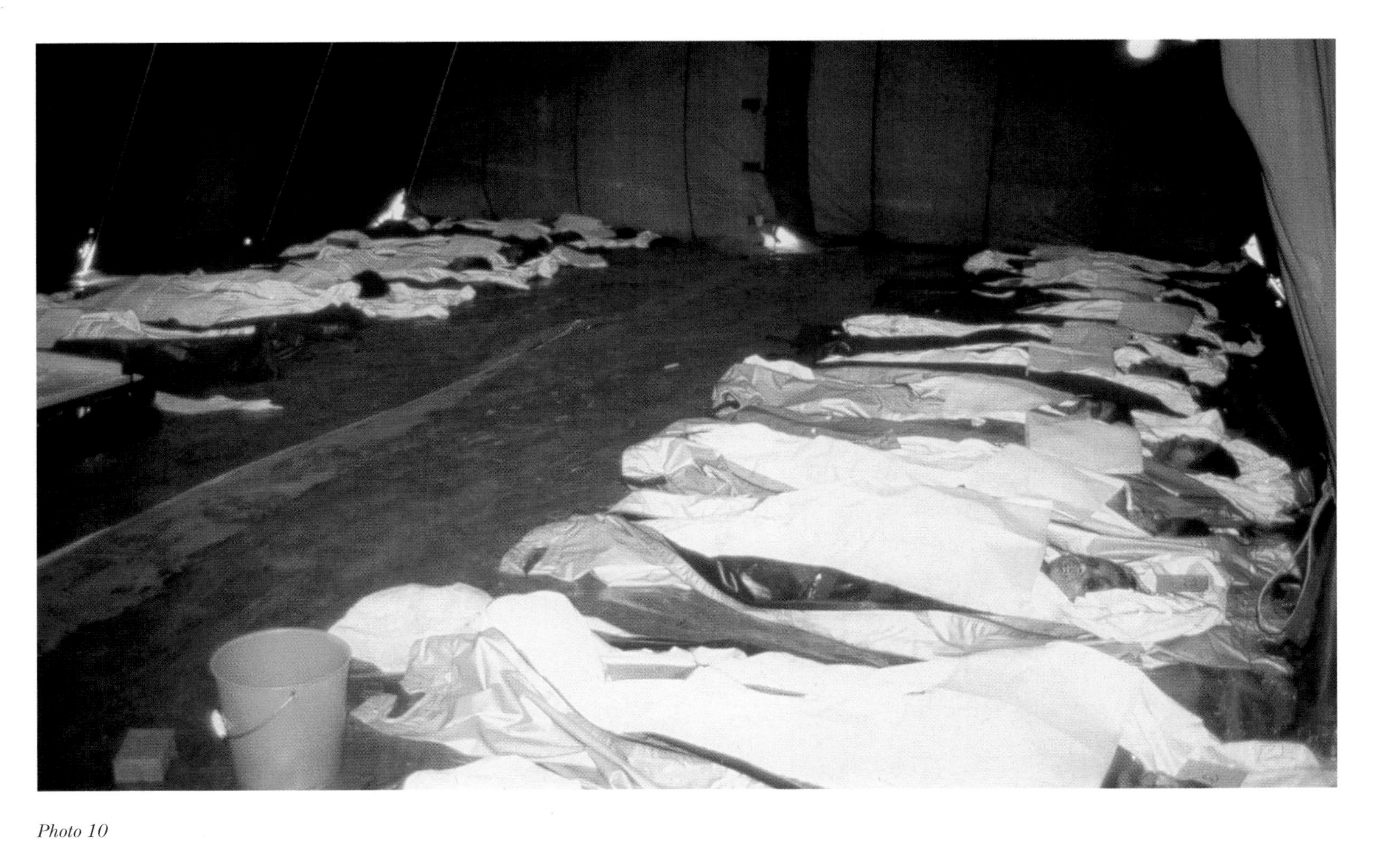

Photo 10

Bodies should not be placed on display for families to identify.

(Photograph: J. Levinson)

Photo 11

Crash of Air Inter near Strasbourg, France. It took several hours before the wreckage was found on a wooded mountain.

(Photograph: Dr Michel Evenot)

Photo 12

Lufthansa crash at Warsaw Airport. Fire fighters removed the black box from the aircraft.

(Photograph: Lufthansa)

Photo 13

A week later firefighters were still at the scene.

(Photograph: Lufthansa)

Photo 14

Putting on protective clothing before handling deceased.

(Photograph: N. Goodman)

Photos 15–19

Aftermath of a bus bombing – No. 5 on Dizengoff Street, Tel Aviv (Israel), 19 October 1995.

(Photographs: courtesy of the Israel Police.)

Photo 16

Photo 17

Photo 18

Photo 19

CHAPTER 13

DAMAGE TO BUILDINGS

INTRODUCTION

As has been mentioned, sometimes buildings are struck in transportation disasters. The need to evacuate can be extremely obvious. Sometimes the buildings are condemned. Sometimes the evacuation is long term, as a building is being repaired. Each type of building – residence, hotel, factory – can raise its own unique problems.

WHEN BUILDINGS ARE STRUCK

The Air France Concorde crashed into a hotel outside Paris in 2000. Residents had to be re-housed quickly. A complicating factor in this case was notifying families far from the disaster scene.

> On 25 July 2000 an Air France Concorde jet en route from Paris to New York crashed into a hotel outside in Gonesse near Paris. All 109 people aboard and four more people on the ground were killed.

In 1987 in Wayne Township, Indiana (USA) a small plane also crashed into a hotel. The pilot of the military craft ejected safely, but nine employees of the hotel were killed. The hotel was well built with numerous fire-safety features (International Conference of Building Officials Code[1] with fire-resistant materials) that helped minimize damage and injury. The rapid arrival of the (primarily volunteer) fire department[2] one minute after aircraft impact was another decisive factor in the response. Also adding to the success of the operation were frequent aviation-related emergency exercises, including discussions the previous day about a forthcoming exercise.

Curiously, the number of guests in the hotel at the time of the incident was known, but the exact number of hotel employees in the building was not known with certainty. (Usually the opposite is expected.) This complicated the effort to

[1] Building code revised and published every three years. Available from Techstreet, 1327 Jones Drive, Ann Arbor, MI 48105, USA (Mention is for information purposes only and does not imply recommendation or non-recommendation.)

[2] The Wayne Township Volunteer Fire Department was later backed up by three other fire departments according to a mutual aid pact that was put into effect.

account for all possible victims, and made a second search of the building necessary (US Fire Administration, 1987).

At 0911 on 20 October 1987 the pilot of an A-7D Corsair single-engine military aircraft notified the Indianapolis, Indiana (USA) control tower of an emergency situation. The pilot ejected, and his plane crashed into a nearby Ramada Inn Hotel.

Another example is an El Al 747 in cargo configuration that crashed into a housing complex in the Bijlmer area of Amsterdam (Netherlands). Many of the building's residents were living in the Netherlands without proper visas and residence permits, hence it was extremely difficult to handle Search and Rescue in the response phase, then insurance and compensation payments after life-saving efforts were completed. Government offers to grant legal status to those affected by the crash encouraged many people to file false claims.

On 4 October 1992 an El Al cargo flight (Boeing 747-258F) from Amsterdam to Tel Aviv (Israel) developed a problem in its no. 3 engine. A pylon separated from the wing, colliding with the no. 4 engine. This caused the no. 4 engine and the pylon to separate. The plane crashed into a residential building killing the four aboard and some 43 people in the building, which also suffered structural damage.

Crashes into buildings on the ground are perhaps more common than generally thought. After all, the aircraft accounts for the large majority of the victims and certainly receives the focus of attention. When Pan Am 103 crashed in Lockerbie, 11 people died in buildings in a residential neighborhood; these victims are often confused in casual parlance with those on the aircraft. There are numerous other examples of deaths on the ground. The crash of Alliance Air in Patna, India destroyed two houses and killed five people on the ground.

Crashes into buildings are not restricted to any one mode of transportation or to any single type of building. There are numerous instances of cars and trucks hitting buildings. These accidents include even trains. No building is immune: on 28 July 1945 a US.Army B-25 bomber crashed into the 78th and 79th floors of the Empire State Building in Manhattan.

On 17 July 2000 an Alliance Flight No. 7412 (Boeing 737-200) was flying an internal flight in India from Calcutta to Delhi via Patna and Lucknow. The plane was on approach to Patna Airport, when it crashed. Forty-nine of 52 passengers and all six crew members were killed. Two houses in a State Government Colony were destroyed, five residents were killed, and another five were seriously injured.

On 6 February 2000 a train from Amsterdam to Basle (Switzerland) derailed and struck a house at Bruhl, Germany (near Cologne), killing at least nine persons. Parts of the train landed in the living room and front garden of the house.

Sometimes, however, following an accident or disaster a total evacuation of an area cannot be effected, even when that evacuation is justified. After the Cubana accident in Guatemala, there was a fire hazard to the neighborhood where the plane crashed. Although there was an effort to evacuate, it was met with only limited success, primarily due to residents' fears for their property.

On 21 December 1999 a Cubana DC 10-30 jet crashed into the La Libertad neighborhood located at the end of the landing strip at Guatemala City–La Aurora Airport. At least 26 people, including nine on the ground, were killed when the aircraft overshot the runway and struck the residential area. The airplane had been carrying 314 people, including 18 crew members; in total over 70 people, from both the aircraft and the residential area, were injured. (Nor was this the first airplane to crash into the same neighborhood! Yet, the neighborhood [reportedly a shantytown of substandard and "privately constructed" buildings] still remains at the end of the runway.)

This was not the first time a Cubana plane struck a residential area. On 3 September 1989, shortly after takeoff, a Cubana Ilyushin 62M fell near Havana, Cuba. The aircraft crashed about two kilometers from the airport. All 126 persons aboard the aircraft were killed as were 14 people on the ground.

Precautionary measures were taken for those who remained. Electricity for the entire area was cut off, and no night-time salvage was permitted, because of the fear that generators might spark fires from spilt fuel.

Guatemala City is not the only example of the poor fearing for their property. The threat of theft or looting after evacuation plays a considerable role in decision making in many poorer areas. In one incident in Bangladesh subsequent surveys showed that between 33 and 50 per cent of local residents failed to evacuate because of fear of theft. In some cases there was partial evacuation, sending women and children to safer ground during times of flooding and leaving men behind to guard property.

Returning home to a building that has been hit by a falling airplane or by a vehicle means that a building engineer has certified the building to be safe. And sometimes it is simply impossible to return home![3]

[3] In the early 1950s there were three air crashes in Elizabeth, New Jersey (USA) in less than a two-year period. Some homes were damaged beyond repair and were leveled. In the 1990s the roof of one of the buildings hit in one of the crashes fell. One can only wonder if that was not due to undetected structural damage.

CONCLUSION

There are numerous examples of aircraft crashing into buildings on the ground. Less dramatic, of course, are cars, trucks and buses that also have crashed into buildings. This adds a second dimension to disasters. It means that search and rescue must be done not only in the aircraft/vehicle, but also in the building. When an information bureau is set up to track the whereabouts and condition of all involved, those victims in buildings must also be listed.

CHAPTER 14

INFORMATION CENTERS

INTRODUCTION

After a disaster it is critical to establish the whereabouts of all persons potentially involved and to answer inquiries from the public.

Many disasters cause displacement for a large number of people. A rush-hour train or subway crash affects hundreds, if not more. There is also uncertainty as to who was, in fact, involved. Hence, there are numerous inquiries regarding people who were thought, perhaps, to have been in the area of the disaster, but in fact they were not involved.

PUBLIC INQUIRIES

Public inquiries are common, but in numerous cases it is impossible to determine the number of telephone calls due to overloaded circuits.

Such incidents as the crash of Pan Am 103 and the sinking of the ferry at Zeebrugge (Belgium) generated at least 30,000 and 40,000 telephone inquiry calls from the public in a matter of days.

In many jurisdictions this is called a *casualty bureau,*[1] but in fact this is a misnomer. Although emphasis might be placed on the injured and dead, it is also important to locate evacuees and advise family that they are safe. For this reason *information center* is a preferred term.

[1] It is the policy of many information bureaux to notify only of uninjured or injured. Notification of death is not made by answers to telephone queries.

THE INFORMATION CENTER

The tasks of the information center are to:

- Receive reports of missing persons
  - Compile lists of missing
  - Record contact information for those making queries
- Receive reports from mortuaries, hospitals and evacuation centers
  - Compile lists of people and their location
- Collate the above two lists

- Make notifications as appropriate
  - Directly to families
- For other agencies[2]
  - Answer public inquiries

[2] To relieve the pressure exerted by relatives on hospitals. This is most effective for concerned friends and relatives who are distant from the disaster site.

In many cases a network of liaison officers, in mortuaries, hospitals and evacuation centers, is required to gather the information required. Assignment of the information center is usually to the police or to the local municipality. (When functioning as a casualty bureau, the task is usually given to the police; when functioning as a wider information bureau with evacuee data, local government plays a more prominent role.) More important than to whom the responsibility is given is that the establishment of the center be preplanned with staffing from relevant offices, including public relations and media liaison.

When the information center receives public inquiries, it reduces the communications load on operational offices (such as hospitals). When people can call one number rather than a series of hospitals, there is reduced load on area telephone exchanges. The success of the information center is, of course, based upon collecting and disseminating accurate information on a timely basis (often judged by public sentiment and not necessarily by objective criteria).

Before the information center is opened for inquiries from the public, the following steps should be taken:

- Brief staff on what to say and what *not* to say
- Keep staff up-to-date on developments and new information
- Review use of forms to record information
- Publicize telephone number
- Provide a reasonable number of phone lines

CONCLUSION

The operation of an information center is the primary method by which authorities can keep track of victim's locations and respond to public inquiries. This information is also important in post-incident reconstruction of what happened.

CHAPTER 15

TERRORISM AND CRIMINALITY

INTRODUCTION

In very legalistic terms, "terrorism" is not a crime. It is a motive for a series of crimes such as murder, attempted murder, etc. It is an attempted explanation or justification for criminality. Since terrorism is a motive and not a type of activity, it can include a variety of disasters such as bus bombing, skyjacking, airplane explosion and arson.

EXAMPLES

The essential difference between a usual criminal action and a "terrorist" incident is popular perception. It is the public that accepts an incident as a terrorist act rather than as a solely criminal doing. Sometimes it is, ironically, police deployment that fuels that perception. The scale of police deployment can also mark the difference between "accident" and "disaster," although the reason for the scale of deployment can have little to do with damage done.

On 17 October 2000 the derailment of eight of the eleven cars of a train in Hertfordshire, England, left four dead and over thirty injured. Although there were no obvious signs of bomb explosion, the police dispatched their anti-terrorism squad in view of a bomb warning placed two days earlier. Even though the anti-terrorism unit was part of the investigatory process, its presence encouraged a popular conclusion of a definite terrorist action.

TERRORISM

From a strictly legal perspective the purported justification of a criminal act is irrelevant to enforcement and prosecution. Nonetheless, acceptance of an incident as "terrorism" adds political considerations to response and post-incident police deployment. The most obvious implication of a "terrorist" act rather than a "criminal" action is the need to restore the disaster site to normalcy as soon as possible to counter the terrorist's perceived desire to disrupt usual behavior patterns. In some cases this has been done before police even finished evidence collection.

It is rare that after a criminal act people take to the streets to decry the breakdown of law and order, yet after a terrorist action those demonstrations are quite common, especially after "neighborhood" incidents such as bombs exploding on local buses or when ethnic populations are involved, a situation which creates an intimacy amongs victims' families.[1] This has the effect of adding an additional drain on police manpower as forces are assigned to the demonstrations.

[1] For example, the bombing on 1 June 2001 of a pub in Tel Aviv with a primarily Russian clientele.

There is no shortage of examples in which the results of a terrorist act have clearly reached the proportions of a full disaster.

Many aviation accidents are caused unintentionally by a variety of technical failures or human mistakes. There have been, however, numerous intentional aviation disasters caused by terrorists. The explosion and crash of Pan Am 103 is only one single example. Terrorist actions have been committed by people aboard the flight, by people under the flight path armed with a launching device, and by people responsible for the placement of a terrorist device aboard a flight, either unaccompanied or taken (wittingly or not) by a passenger.

CRIMINALITY

Criminal acts have included skyjacking for ransom money and even shooting of the pilot and pilot suicide. The latter was reportedly the case for the Royal Air Maroc flight which crashed near Agadir, Morocco on 21 August 1994. The disaster response was affected by the circumstances of the cause; Moroccan officials delayed the arrival of foreign disaster victim identification teams that wanted to examine the bodies of the victims.

DEFINITIONS

It is very difficult to define what constitutes a terrorist attack against transportation. In all probability that definition should include not only acts against means of transportation (buses, airplanes), but also acts against the logistical support facilities, such as the bombings of LaGuardia–New York (29 December 1975), Madras (3 August 1984) and Frankfurt (1985) airport terminals.[2]

[2] Not all bombs explode. A terrorist was arrested on 14 December 1999 with 130 lbs of explosives reportedly to be detonated at Los Angeles Airport (LAX) on 1 January 2000. The person was convicted in court on 6 April 2001.

The LaGuardia Airport bomb exploded in the terminal building, killing 11 and wounding more than 80. The motive is presumed to be terrorist-related. To this day, however, no arrests have been made in the case.

A bomb linked to Sri Lankan militants (Tamil Ealam Army) exploded at Meenambakkam Airport (Madras, India), killing 30 people. The bomb apparently contained gelatine sticks, a detonator, perhaps red phosphorous and sodium metal, and potassium cyanide (Jain, 1997).

SERIAL CRIME

Another problem in defining terrorist incidents as terrorism is that there can be a series of attacks launched by one group, yet not all of the attacks are against transportation. This is serial crime in its most classic sense. The 27 June 1994 release of sarin gas in Matsumoto, Japan (150 kilometers to the west of Tokyo) had no apparent transportation connection. On 20 March 1995 the same perpetrators released sarin in the Tokyo subway system. Matsumoto was evidently a trial for the Tokyo incident. The lesson learnt is that a broad perspective must be taken, taking into account that transportation can be just one of the targets of a terrorist group. Phrased differently, the fact that a terrorist group did not target transportation *this* time does not mean that it will not do so *next* time.

Chemical agents are grouped as "G" (volatile) and "V" (non-volatile). Primary nerve gases are GA (tabun – $C_5H_{11}N_2O_2P$) and GB (sarin – $C_4H_{10}FPO_2$). GD (Soman) and VX are more effective, but they are also harder to produce. The Germans developed sarin during the Second World War; its name derives from the initials of the Nazi scientists who worked on the project.

The Tokyo incident caused 12 deaths and thousands of injuries. Several devices had been placed in the subway system during morning rush hour, but most malfunctioned. Only two devices released gas.

On 19 April 1995 more than 400 subway riders in Yokohama, Japan were dispatched to hospital following a gas release in the subway. On 5 May 1995 yet another attempt was made to release gas in the Tokyo subway system. This time a cleaning worker discovered a canister of hydrogen cyanide that was set to explode. On 6 July 1995 there was another incident in the Tokyo subway system.

The sarin incidents at first caught the Japanese unprepared, particularly with regard to the large numbers of people involved. There was also secondary contamination of hospital staff, with 23 workers showing miosis and complaining of nose and throat congestion. When dealing with possible nerve gas ingestion, decontamination and ventilation of treatment facilities are mandatory.

In the field, protective clothing and equipment (PPE) against sarin (and other nerve gases) include totally non-permeable covering (headgear, clothing, gloves, boots). The net effect is an inability to deal with perspiration, even with a breathing system available. Destruction of the clothing after use is advised.

Decontamination of affected persons must be immediate, particularly if sarin liquid touches the skin. At least one medical authority has stated that time is more essential than deciding which decontaminating agent is best.

CRIMINAL NEGLIGENCE

Regulations and measures to prevent disasters, or at least to mitigate their consequences, are of no use unless they are obeyed.

> On 16 September 1991 a fire broke out on the *Omnisea*, a 324-foot seafood processing ship that was docked in Seattle, Washington (USA) Harbor. Fire caused damage estimated between $7–10 million. The response was complicated by the presence of Hazmat (ammonia – needed for the refrigerator system to cool fish) stored aboard.

At the time of the fire, after a cutting and welding procedure needed for repair, the ship's crew was reportedly on coffee break with no one on fire watch.[3] This was in violation of the Seattle Fire Code and constitutes a criminal violation that can lead to charges of criminal negligence (US Fire Administration, 1991).

[3] Verified by Seattle Fire Department Fire Investigator's Scene Report 91-31790.

PROSECUTION

In today's judicial world, legal ramifications are an important consideration that cannot be ignored in a disaster response. Kanarev (2001) suggests a basic dichotomy for legal issues: (1) emergency *powers* for response, and (2) operational *liabilities*.

EMERGENCY POWERS

It is extremely unlikely that special governmental powers (e.g. state of emergency, martial law) will be enacted due to a transportation disaster. These widespread powers are invoked after much larger disasters, such as earthquake. These powers enable authorities a much wider range of activity, such as entering property without permission (e.g. to deal with utility lines) or to effect mandatory evacuation.

OPERATIONAL LIABILITIES

Many disasters are followed by legal actions. In *criminal* court there is society's need to find a villain – the person or organization responsible for the disaster – and to punish those responsible for the consequences of the disaster.

As a corollary, *civil* suits have also become an increasingly common course of action in today's adjudicative society. The inclination to initial court proceedings is all the more present after disaster, when people feel that legal actions are one method of recovering damages or perceived damages, and making amelioration for suffering or perceived suffering. Once, the families of deceased persons quietly accepted compensation offers from airlines. Today the trend in many circles is to challenge offers.

DEFINING THE OFFENSE

A primary prerequisite to prosecution is definition of an offense. That requires a sound legal analysis of the disaster. That analysis must take into consideration legal responsibility and operational functioning. The mere fact that someone was not able to fulfill his legal responsibility does not mean that he is necessarily liable. Plausible actions and restraints, mitigating circumstances, limited information available, time constraints, and other factors can provide adequate defense.

PROBLEMS IN PROSECUTION

Prosecution is not simple and, particularly in aviation and maritime-related cases, court procedures can take several years.

An offense committed in the sovereign jurisdiction of a state and leading to a disaster is prosecuted according to the laws of that state. The matter, however, becomes more complex when talking about a crime over international waters. In aviation voids of prosecution authority are adjudicated by the United Nations International Civil Aviation Organization (established 1947), with appeals to an ad hoc appeals board, or the International Court of Justice (Vermylen, 1991).

A recent summary of legal arguments related to aviation can be found in United States District Court for the Eastern District of Pennsylvania MDL No. 1269, re Air Crash (Swissair) near Peggy's Cove, Nova Scotia on 2 September 1998.

CONCLUSION

In recent years a number of law firms have begun to specialize in transportation disasters (primarily maritime and aviation). Their experience can be an effective resource in legal action.

SECTION III

TECHNICAL ASPECTS OF A RESPONSE

Chapter 16
COMMUNICATIONS AND INFORMATION

Chapter 17
MEDICAL RESPONSE

Chapter 18
POLICE OPERATIONS

Chapter 19
FIRE

Chapter 20
FORENSIC SCIENCE

Chapter 21
VOLUNTEERISM

Chapter 22
PSYCHOLOGICAL TRAUMA AND MENTAL HEALTH

Chapter 23
MEDIA AND INFORMATION

CHAPTER 16

COMMUNICATIONS AND INFORMATION

INTRODUCTION

It might seem unusual to start a discussion about handling a disaster by raising the topic of communications. It is, however, communications that can dictate the success or failure of a disaster response. If responders are not in contact with each other, and if information (whether it be reports or instructions) does not flow properly, it is hard to envision a successful disaster response. This is true for all types of disasters. It is just as true for a plane crash as for an earthquake.

COMMUNICATIONS LINES

It seems that there are never enough communications lines available in a disaster response. If, proverbially, work expands to fill available time, then communications expand to fill available lines or channels.

Use of communications lines can be prioritized. In one disaster response, available telephone lines were divided amongst responding agencies, each of which fixed "quotas" and time limits for incoming and outgoing calls.

There is also a public need for communications. If it is hard to prioritize those communications, a time limit can certainly be imposed on each connection. In an earthquake, for example, people are inclined to make a high proportion of very short calls to assure out-of-area friends and relatives of their safety; in a transportation disaster telephone lines are also needed for the same purpose.

There are basic differences when considering what type of disaster has occurred. In natural disasters a significant segment, if not the entire communications infrastructure, can be disrupted or damaged. In a transportation disaster, however, the basic problem tends to be overload (that is not to say that specific equipment cannot be damaged or even destroyed). There is a more structured environment, in which authorized government or commercial sources can specify use of particular telephones with override priority to those lines.

There is no way to control availability of cellular phone lines for the convenience of disaster victims. If victims do not succeed in using their own cellular phones due to overload, a very fair solution is to make other telephones available.

In a disaster it is not only the number of lines that is needed. Very often special service is required, such as conferencing. These services should be coordinated in advance with the telephone service provider as part of general disaster planning.

If communications are used properly, a significant amount of information is *received* from outside sources. This is particularly important in disasters. The general rule is confusion in initial stages due to the *lack* of confirmed information. When factual reporting does become available (whether it be from the scene, from hospitals, or from other sources), there must be a communications infrastructure to relay that information to persons needing it. Again there can be a communications overload, not only in terms of telephone lines, but also regarding the volume of information supplied. In later stages following a disaster there can be so much information available that prioritization can be necessary. Senior response authorities, for example, need concise information giving a general overview, and not a constant stream of operational details.

CONCLUSION

Adequate communications are vital for disaster response. Lines are easily overburdened, and there must be preparation to give essential users, such as police, fire and medical, priority use.

CHAPTER 17

MEDICAL RESPONSE

INTRODUCTION

Medical response is planned in accordance with cultural norms and resource availability (usually a resultant factor of cultural demands). In the First World War certain armies had ten veterinarians for every doctor. This was based upon "horse-centered planning," valuing the horse as a weapon of war. Although today this might seem to be an extreme example, there are similar priority judgements made in modern medical planning. The harsh reality is that medicine is expensive, and life *does* have its price. If this were not the case and finances were not a consideration in saving lives, we would all drive vehicles akin to tanks that would tremendously reduce serious injuries and fatalities in road accidents.

TRIAGE

In general medicine the role of a physician is to save lives. In disaster medicine the role of the physician is to do the most good for the greatest number of people. This is a very different approach, often causing medical personnel serious psychological problems. Taken to the extreme, there are times in disaster situations when doctors must knowingly allow the seriously injured to die (even though they might have been saved under other circumstances), so that less seriously wounded patients can be treated (and not become more serious).

At a disaster site it is the job of medical personnel to prioritize the evacuation (*triage*) of injured persons and to certify the death of deceased victims (Peschel and Peschel, 1983). This is usually accompanied by a Medical Emergency Triage Tag.[1] Such tags, available in several commercial versions as well as government-produced printings, have quick methods of recording patient condition (from lightly wounded through fatality), and sometimes give the option of making medical notations.

In theory, triage enables evacuation in the order of medical need. An accompanying system benefit is to reduce the number of high priority evacuations,

[1] Terminology of the American Civil Defense Association, which markets a suggested tag. (Mention does not imply endorsement or recommendation of a product.)

hence lessening pressure on the disaster response mechanism. Another by-product is the theoretical analysis of medical need, and the sending of the injured to the most *appropriate* hospital (not necessarily the closest) based upon treatment considerations and available space.

Search and Rescue personnel should be instructed to move patients to triage rather than directly to ambulances. In an airport situation this is often an unnecessary worry, because all SAR staff are usually airport employees who *should* be familiar with procedures. An airport, as a closed area, enables considerably more incident control. Instructing initial SAR responders is more of a concern in uncontrolled areas where personnel without prior experience working together are called upon to respond.

In reality, triage is most effective when the number of injured persons outnumbers available ambulances. At the moment when the opposite is true, the most common *de facto* method of evacuation is simply "scoop and run" – that is, take patients and transport them to hospital. (Even in this situation some EMS [Emergency Management System] responders may still adhere to the illusion that they have done triage by "selecting" an injured person to be sent to hospital). This has the adverse effect of transferring triage to the receiving hospital emergency rooms, where examination and admission decisions must be taken without the benefit of previous field prioritization. Although input can be obtained from examination in the ambulance, that information tends to be limited and does not place the patient within the context of other injured persons.

This procedure assumes professional considerations. In many cases the public works under the unrealistic expectation that all persons must be brought to hospital as soon as possible.

TRANSFER TO HOSPITAL

Another common misunderstanding is that triage is a purely medical treatment activity. One of the key functions in triage is non-medical. It is maintaining logistical communications with area hospitals to ascertain patient load and bed space, so that the injured can be dispatched properly. A corollary of the function is to alert hospitals to the patient load dispatched and to unique problems so that staff can prepare accordingly.

In urban areas experiencing moderate-size disasters it is not unusual to find that arriving ambulances, in fact, do outnumber the wounded. Another contributing factor is that sometimes people must be extricated from trapped situations, meaning that immediate transport is not possible, thus there is more time for additional ambulances to arrive.

Current medical thinking is that in serious injury cases, basic stabilization should be attempted before ambulance transport. In a mass emergency the ratio of medical responders to injured does not always allow stabilization in the field (e.g. CPR, endotracheal intubation, Advanced Cardiac Life Support) (US Fire Administration, 1994).

It is prudent to examine even the apparently uninjured for possible unseen or internal injury – in other words, the examination is to discover any possible injury not detected by the victim himself. Such an examination can also be a useful tool in counteracting any exaggerated or fraudulent claims filed later.

The method of removal to hospital is also a medical decision; in certain circumstances it is not desirable to fly a victim by helicopter, even though that might be the quickest evacuation method. Air evacuation must be reserved for selected cases, since in a large event the number of helicopters available and landing space can impose severe practical limitations.

Very often medical personnel are not in control of the evacuation process, and it is all too frequent that the *less seriously injured* persons evacuate themselves to hospital. Following the 1949 *Noronic* ship fire in Toronto, there was no control of movement to hospital, and many people turned up on their own initiative. Only 45 of the first 136 people arriving at hospital required admission (T. Brown, 1952).

CASE HISTORIES

On Saturday, 11 May 1985 a fire broke out in the grandstand of the Bradford (England) soccer stadium. Within two minutes virtually the entire grandstand was engulfed in flames. Thirty people with minor injuries arrived in hospital emergency rooms before ambulances brought in the seriously wounded. A general breakdown in communications coordination followed, although each hospital worked well as an independent operation. In all, 53 soccer fans died, and 280 needed medical treatment (Scanlon, 1991).

In November 2000 a bomb exploded next to a bus in Hadera, Israel. In this case the least seriously injured also evacuated themselves to hospital, but in stages after the incident. Twenty-seven injured people were evacuated by ambulance within an hour of the explosion. Within the next four hours a similar number of lightly injured people also found their way to hospital.

BODY REMOVAL

Sometimes taking the dead comes under the heading of evacuation, but an alternative term often used is *body removal*. No matter which term is used, every effort should be made not to move the dead to the triage area. (This is problematic in actual practice, since in many cases removal of the injured from a stricken area is done by persons not authorized to certify death. Therefore, all intact bodies are moved to triage.) Nor should a temporary mortuary be established in a hospital. During a disaster triage areas and hospitals will be busy enough dealing with the living; there is no reason to increase that burden with the dead. Thus, under no circumstance should those victims certified as dead be moved to the hospital or the hospital mortuary; it is much better to use institutes of forensic medicine, funeral parlors, or makeshift facilities.

When bodies are removed from a disaster scene, they should be placed in sealed body bags to prevent loss of body fluids. Body handlers should also be instructed not to presume ownership of property and instinctively include nearby items in a body bag. Sloan (1995) shows conclusively that associated property can be in error, hence as a safeguard all property should be checked upon opening of the body bag in the mortuary.

Local religious and cultural attitudes towards the dead must be taken into consideration. In Hindu and Moslem areas there is a religious obligation of rapid cremation (Hindu) or burial (Moslem), sometimes shortcutting body registration and formal identification, much to the consternation of Western observers.

PARAMEDICS

As a general rule ambulance services function relatively well in a disaster environment, since paramedics are performing their standard function. Routine coordination exists between police and fire services, so relatively few problems are experienced in this area . . . until non-local forces are brought in, and daily working relationships are no longer effective.

To take ambulance workers away from the routine tasks they do best can become problematic. Difficulties arise in large disasters when there is a temptation to transform ambulance services from first aid, stabilization (both are to be considered short-term treatment), and evacuation into a replacement for hospitals to deal with the less seriously injured. This is a distortion of roles.

HOSPITAL RECEPTION AND RECORDING

Receiving a large number of patients following a disaster requires implementation of a hospital emergency plan that insures sufficient staffing, information services, clearing of required treatment areas, and opening of waiting rooms for families. This emergency plan must be coordinated with general regional planning, so that as far as possible there be a smooth flow of injured.[2]

Following the Manchester (England) Airport disaster of 22 August 1985 at 0713, area hospitals were placed on standby (0722). At Wythenshawe Hospital, when the first bus arrived with 36 victims, the hospital itself decided to upgrade

[2] The American Joint Commission on Accreditation of Hospital Organizations (JACHO) requires the hospitals to exercise their own emergency plans semiannually (Aghababian, 1986).

A Boeing 737 caught fire on the runway at Manchester Airport on a holiday flight bound for Tenerife (Canary Islands). Fifty-five people aboard the plane were killed.

operations from standby to emergency status. Official upgrade notice was not received. The last ambulance arrived at the hospital at 0845, yet only at 0956 was stand-down notice received from ambulance control. Although emergency systems are usually thought of in terms of pressure on staff and bed space, Wythenshawe also experienced very heavy demand on the hospital pharmacy, and of course on the communications system. One part of the hospital emergency plan that worked quite well was the posting of signs directing people unfamiliar with the building to the various emergency facilities.

Unless a hospital has direct communications with the on-site triage dispatcher, it has no way of knowing if and when critical patients are to arrive. At what point staff can start treating the lesser wounded is also not usually clear.

A hospital/dispatcher two-way communications system is an important input into logical patient distribution amongst area medical facilities. Hospital communications, however, cannot only be with the triage and/or ambulance dispatcher. What is also needed is an intra-hospital and ambulance communications network to adjust hospital selection for patients based upon diagnosed medical conditions.

Selection of receiving hospital is not always made on a basis of medical considerations or physical proximity. In the M1 air crash, initial dispatch of wounded after the crash was done on the basis of traffic problems and the probability of an open road (Smith, 1992). For the record, in this incident there was a communications breakdown. Area hospitals started receiving patients before official notice was given of the crash.

A Boeing 737-400 took off from Belfast, Northern Ireland headed for London–Heathrow at 1952 on 8 January 1989. The pilot declared an emergency after vibrations shook the aircraft, and smoke filled the cabin. He tried to land at East Midlands Airport. The plane crashed at 2024.43 on the M1 highway embankment at Kegworth (England); 339 died immediately, 8 died of injuries, and 79 survived (74 of these with serious injuries).

As the injured are received in hospital it is important to register them, not only listing injuries, but also recording the area from which they were evacuated. In an exercise at Ben Gurion Airport near Tel Aviv victims from an "air crash" were evacuated to hospital. At the same time ostensible victims from a "traffic accident" were also moved to the same hospitals. It soon became clear that considerable work was necessary to determine which patients belonged to which accident. This complicated the reconstruction of the airline passenger list with location of all victims.

Another problem in constructing lists of disaster victims and their location is that sometimes hospital records, even to the minimal level of patient name, cannot be released without legal consultation due to patient privacy laws.

Although the emphasis in the treatment of victims must always be on health concerns, attention should be paid to saving all of the patients' property, even if items do not appear to have any significant financial value. Even property such as ripped clothing can be needed for incident reconstruction and victim identification.

It is necessary to station a liaison officer in each hospital to which the injured might be brought. It is his or her task to report to the incident command center or EOC (Emergency Operations Center) the names of all victims arriving in hospital, whether admitted, under treatment, or just being examined. This is necessary to keep accurate records for any inquiries received. In closed population incidents this reporting gives responders an accurate accounting of who is still missing.

HOSPITAL INFORMATION CENTER

Drory *et al.* (1998) proposed a scheme for a hospital information center to handle public inquiries, both in person and by telephone. The goals he defined are to:

- provide reliable information systematically collected;
- help members of the public cope with uncertainty;
- relieve medical staff of the responsibility of dealing with the public.

Although this suggested program, seemingly based upon the experience of a Tel Aviv hospital, is well intended, it misses the primary goal that, particularly in a metropolitan area where there are numerous hospitals, an information center should be a regional effort. That is much more effective than each facility fending for itself and tying up personnel. It should be noted that even the Municipality of Tel Aviv is trying to establish an area-wide information bureau (including several municipal jurisdictions) to handle disaster situations.

A hospital-based information system also does not take into consideration victims who are treated at temporary facilities for examination and light injury.

The goal of relieving doctors of the responsibility to provide information is also questionable. An information center might lessen their administrative load, but it certainly will not eliminate it, since the most accurate patient status reports will still come from the medical staff.

Locating all disaster victims evacuated to hospital is not necessarily an easy task. It can take time, even when preplanning does exist, because of the extensive logistics involved. Following the Avianca crash in Long Island, 71 people were transported by ambulance to 17 different area hospitals, then another 22 were brought to hospital by helicopter (in good part due to congestion on the access road to the crash site) (Meade, 1990).

Some medical experts have suggested that disaster victims should occupy no more than 20 per cent of a hospital's bed space. That is a wishful goal beyond the hospitals' ability to control.

MEDIA

One factor that is hard to control in a disaster scenario is the arrival of the press at hospitals. Within 24 hours of the Bradford fire, the local hospital was inundated by more than 100 press corps members. When politicians visited the injured, this number only increased. Part of dealing with this is obviously preplanning, including an information release schedule which will ostensibly reduce the number of media members between scheduled press conferences.

As has been described, the activities in a hospital as it responds to disaster are complex, often including functions that are not included in routine work. Just as in any other disaster function, the hospital must have a written disaster response plan that is exercised periodically. Those exercises should be open to the media, so that they are acquainted with the disaster response process in advance of a disaster.

It is insufficient that each area hospital have only its own disaster plan. The plans of each of the area hospitals must be coordinated, so that the hospitals all work together to provide the best medical care possible under the circumstances.[3]

[3] Those hospitals should also have a plan to receive patients if there is an evacuation from another medical facility.

DISEASE AND PROTECTIVE CLOTHING/EQUIPMENT

Usually, perceived health problems are connected to disasters such as earthquakes in which it takes considerable time to extricate bodies. This concern, however, also exists with air disasters that occur in desert or jungle conditions, where even finding the wreckage can be problematic. One example is the crash in the Congo of an internal Buz Air flight to Kisangani in June 1997. The wreckage was found quickly; however extremely poor storage conditions in high temperatures allowed the 27 bodies of those aboard to deteriorate very rapidly.

A myth has developed that there is a risk of cholera and typhoid when bodies are left exposed to the elements. De Goyet (1999) has shown that this fear is unfounded, unless the victims, themselves, were disease carriers – in which case they probably constituted a greater health hazard when they were alive! De Goyet suggests that efforts should be placed, instead, on sanitation, food and water conditions.

There are situations in which the natural immunity system of a person combats any existing germs/bacteria. When the person passes away, and immunity systems ceases to function, those germs then multiply and eventually spread. If major plagues are unlikely, localized disease is a definite and practical worry – due to exposure to infection from victims' body fluids. Those worries include both viral and bacterial infection. All disaster workers should wear protective clothing. There is, however, no universal standard. The type of clothing very much varies according to the incident.

Protective clothing and equipment (PPE) is divided into four levels:

- *Level A:* Radiology. Biology. Highest protection for respiration, skin, eyes.
- *Level B:* Chemicals. Highest protection for respiration. Lesser for skin.
- *Level C:* Chemicals. Air-purifying respirator, moderate skin protection.
- *Level D:* General. No air purification. Moderate skin protection.

One concern is sharp edges on broken materials (when present necessitating heavy gloves, hard hat, and steel toe shoes [rubber/plastic in water]).

Open wounds of responders can result in possible exposure to any diseases that might be carried by disaster victims. All body fluids of victims, even something as common as saliva dripping from the mouth, also constitute potential health hazards for responders.

Hot metal after a fire requires yet another type of protection. In appropriate cases clothing should also be designed to protect against excess noise (sometimes from extrication equipment) and from outside temperature.

Simple masks covering mouth and nose are usually placebos that only concentrate bad odors in front of the nasal cavity. Surgical masks are designed, in part, as a protection for the patient from germs emanating from the medical staff and vice versa; when dealing with the dead that consideration is invalid. As Hooft *et al.* (1989) describe, these masks lead to "further mystification, discomfort, and hyperventilation." Hooft *et al.* add very pointedly, ". . . the use of such masks only strengthens the old belief in the contagiousness of the deceased and death."

Sophisticated masks with filters are required for many Hazmat responses. Masks are also beneficial in the rare case of powdered body fluids (most often blood) possibly carried by winds and entering the nose or mouth of responders.

There are several different types of sophisticated breathing masks:

- *SCBA:* Self-contained breathing apparatus. Tank backpack.
- *SABA:* Supplied air breathing apparatus. Umbilical cord to additional air supply.
- *CCRS:* Closed-circuit rebreathing systems. Version of SCBA, filtering out carbon dioxide and recycling air.
- *PAPR:* Powered-air purifying respirator.[4] Filter for compressed air supply. Used by military for chemical/biological warfare.

[4] The US Fire Administration recommends that training be conducted before this equipment is used.

Avoiding bad odors (e.g. the stench from cadavers) is an easier problem to solve. One suggested method is to spread something like mint-flavored toothpaste or petroleum jelly under the nasal openings.

The improbability of plague does not mean the improbability of infectious disease. Pre-incident immunization of planned responders to disasters should be coordinated with medical authorities. It is prudent to ascertain that inoculations for tetanus are up to date. Hepatitis is another common concern.

The most common hazard is not at all exotic and sometimes overlooked. At disaster scenes items often penetrate shoes. This is a common and recurrent problem. Responders should, therefore, wear strong shoes (both soles and sides).

DECONTAMINATION

When persons are exposed to contaminants, they must be decontaminated before entering society so they do not spread the Hazmat material with which they came into contact. There are numerous suggestions on wash and rinse solutions, depending on the exact nature of the Hazmat. The decontamination area is called *clean zone entry* or other similar term.

Under certain circumstances high-pressure water and detergent soap can be effective cleansers. What is critical in all cases, however, is that all water or other liquids be collected for special disposal.

Particularly in the case of chemical and biological war materials,[5] time can be a key factor in decontamination. In some cases there can be a period as short as ten minutes in which decontamination is effective.

[5] A key institution for research on this subject is the Chemical and Biological Defense Information Analysis Center at Aberdeen Proving Ground, Maryland (USA).

Equipment and clothing should be decontaminated according to approved procedure or destroyed by appropriate method. Deceased persons should also be decontaminated (both externally and internally if Hazmat has been inhaled). Sealing a deceased person in a bag is only effective if the Hazmat will decompose after a short period of time; otherwise, the seal will eventually break in the grave.

MEDICAL EQUIPMENT

Very often people in an accident lose such essential items as eyeglasses, medical apparatus, or medications, particularly in the wake of a sudden disaster that requires rapid evacuation. In some cases the disaster victims do not know the technical names of their medicines nor appropriate dosages. When doctors' records are also destroyed in the same disaster, this can pose major problems. Victims of a disaster may have to undergo an entire series of tests to determine their needs. This may include medications needed on a regular (and sometimes urgent) basis. Their prescription can be a basic problem of professional responsibility for a doctor unfamiliar with the person's medical file. Hospitalizations, therefore, may be necessary after disaster or evacuation to re-establish medical treatment.

WALKING WOUNDED

After many disasters there are people who appear not to be injured or suffering only from minor scratches and bruises. As has been pointed out, in every case it is proper to examine all people involved in an accident to verify that there are no internal or otherwise undiagnosed injuries. This should be done before these people are released.

However, total control is not always possible. In the Warsaw crash, at least one passenger walked away from the crash and, without notifying anyone, flew off to his destination on another flight. Usually this reaction, ranging from deliberate departure to distance oneself from the unpleasantness of the incident to merely wandering off, is attributed to psychological trauma. This is medically unsound and might well create problems in claiming insurance for any injuries later found, not to speak of the confusion it causes in reconstructing the passenger list and accounting for all persons.

On 14 September 1993 Lufthansa Flight No. 2904 (Airbus 320-211) crashed in Warsaw, Poland. Two persons (one passenger and one crew member) died; 54 others were injured and transported to eight different hospitals. Thirty-six ambulances arrived at the crash site.

The unauthorized departure of this one individual is not an isolated incident, nor are such instances restricted to aviation. On 4 January 2000 two trains collided near Elverum, Norway, some 150 kilometers north of Oslo. Nineteen persons died. In at least one stage of the response operation it was thought that several uninjured train passengers had simply left the area.

COMPUTERIZATION OF VICTIM TRACKING

Many hospitals use computer programs (most often bar code) to track the location of patients and the treatment they receive. In a disaster response program these individual hospital computer programs are best coordinated with a fatalities tracking program and with software at the general information center. This is the easiest way to have up-to-date information about a victim's location and condition.

There are two cardinal questions regarding the operation of such computer programs. One issue is the blocking of information, so that only restricted work stations have access to sensitive personal information. The other issue is how to decide which hospital arrivals are disaster-related and which are not; this is not always obvious, particularly in the initial pandemonium of disaster response.

CONCLUSION

Medical treatment starts at the scene of a disaster and extends to treatment centers (temporary or standard hospitals). Needless to say, contingency medical planning should be coordinated with police, so that traffic patterns can be established for emergency vehicles, and so that patients' whereabouts can be registered in the incident information center. In many instances police investigators will want to question or interrogate victims concerning what transpired.

CHAPTER 18

POLICE OPERATIONS

INTRODUCTION

In many jurisdictions the police play a central if not dominant role in disaster response. This is due not only to a specific police mandate, but also to the fact that as a 24-hour operation, police are always on call.

PREVENTION AND LICENSING

Police operations to prevent disasters are most notably oriented toward terrorism and other similar criminal activity.[1] This fits into the general role of police, and it is, quite clearly, no minor function. It includes explosive detection, intelligence collection, deterrent patrolling, and counter-terrorism operations.

No less important, however, is the less dramatic responsibility to provide input before a scheduled public event (e.g. rock concert, political demonstration) can be licensed. Before any such license can be issued, it must be ascertained from the licensee that no unnecessary hazards or dangers exist; the police must also have their emergency plan prepared just in case something does go wrong.

The granting of that license is often not understood properly. One common misconception is the presence of one or two ambulances at a major public event. The stationing of an ambulance outside a stadium with a major soccer game being played is by no means part of disaster planning. It is part of the standard medical plan to treat isolated incidents that happen to individuals. For example, a single ambulance is quite appropriate to handle an injured sports player or a fan who has fainted. It does not even begin to answer medical needs in a disaster. If a license is issued properly, there will be a plan in place to signal arrival of significant medical backup in case of a true disaster.

[1] Police efforts regarding traffic tend to be taken for granted.

PLANS AND EXERCISES

Many aspects of police work do not need to be exercised, since they are routine and are implemented on a regular basis. Special disaster response activities, however, are an exception to this generalization, since disaster planning is not implemented very often. For that reason planning and exercising those unique functions are essential.

> The explosion in the Nypro Factory, Flixborough (England) on 1 June 1974 at 1653 hours is an example of the lack of a police plan. In all, 28 died and 36 were injured at the factory, and more than 30 additional people were injured and 200 houses left uninhabitable in an adjoining residential neighborhood. Police response was lacking. The local Humberside Police had been created only two months previously. Their emergency plan was still "being written," and it had, obviously, never been exercised. The response included many non-routine police functions.

Response to *every* disaster scenario cannot be planned and certainly not exercised. Plans must be flexible so that they can be applied to a wide variety of scenarios. Exercises must take into account the most common disaster problems with emphasis on developing an "emergency mind-set" – that is to say, a way of thinking and not merely reacting to the written word. An example of this thinking is in the Renfrew and Bute (England) Police, where an air/rail disaster exercise helped in their response to a gas line fracture at Clarkston Toll on 21 October 1971.

It is generally acknowledged that the most common disaster response problems are coordination and communications – two interrelated concepts as noted by auf der Heide (1989). This begs the conclusion that these two facets of response must play a primary role in response exercises. Although internal agency exercises are essential, in practical terms this conclusion means that there must also be inter-agency exercises that stress inter-agency cooperation and communications.

It is relatively simple to exercise cooperation between government agencies, even if some are local and others are regional or national. These agencies are built on a hierarchical structure in which directives are routinely issued. It is much more difficult to run exercises that involve the private sector, which is difficult to organize, and in which time can be at a premium.

INTERNAL AND EXTERNAL POLICE COMMUNICATIONS

Disasters, no matter what their type, always place a strain on communications from police to police and from the public to the police. It is not infrequent that emergency telephone lines are overloaded, just at the time when receipt of telephone calls from the public is most critical. It is a known fact that disasters create a burden on the telephone system as people try to find out what happened.

In 1940 a blast destroyed an explosives factory in Kenvil, New Jersey (USA). At that time with fewer telephones in service, reports recorded 20,000 disaster-related telephone calls in the first 10 hours (*Newark News*, 12 September 1965).

One of the many examples is the El Al crash in the Bijlmer area of Amsterdam, where a ranking police official reported, "I had the use of car telephones . . . both are entirely useless. In no time all everything (sic) became overloaded" (Welten, 1993).

On 2 November 2000 a car bomb exploded near the Machaneh Yehudah marketplace in Central Jerusalem (Israel), leaving "only" two dead. Even in this relatively small incident, police cellular telephone lines were overloaded.[2]

One solution suggested in Great Britain is the diversion of calls to other police forces (the distance of diversion being based upon quantity of calls) where other trained officers can respond after being given a short briefing.

An immediate operational response cannot be given to all citizen complaints during a disaster. It is often necessary to prioritize complaints and respond only to the most urgent. This is a reality that often causes complaints by unwitting citizens. If nothing else, recording of civilian complaints during a disaster give an overall (though not confirmed or critical) situation evaluation.

[2] Another problem often encountered with cellular telephones is battery life. In a protracted response spare batteries with full charge should be kept in reserve. Planning should also take into consideration cellular phone (and radio) "dead zones." This can be particularly acute in handling offshore incidents.

DEPLOYMENT

The most widely used managerial method for deployment in a large operation is the *Incident Command System* (ICS),[3] of which there are several versions, each with its own minor variations and details. This is a civilian version of a military-style command and control hierarchy. In short, the system allows flexibility in expanding and contracting operations as the incident develops. It also allows for the inclusion of auxiliary or outside forces. A basic misunderstanding of ICS is that it is not a quick fix for management at an incident. It is an overall support system that requires advance preparation and training so that it can be activated and used properly as the situation requires. Nor does it replace the agency-specific tactical commands.

[3] Originally developed for fire-fighting forces after two weeks of fires in Southern California in 1970, that affected 580,000 acres and took 16 lives. A not-as-popular alternative to ICS is the National [USA] Inter-Agency Incident Management System (NIIMS) which was used in the Galaxy crash response.

Disaster response can require an extensive workforce, often necessitating the activation of *mutual aid pacts* (pre-arranged agreements between response agencies in different jurisdictions, as opposed to unplanned *ad hoc* assistance).

In Lockerbie, admittedly an extreme case of long-term mutual aid, 5871 policemen were fielded during the first week. A month later the operation still required 1693 policemen. One year later 50 detectives were still working on the case on a full-time basis. Part of the manpower was needed because of the spread of the incident. The search area extended for 845.72 square miles. Bodies (as opposed to airplane parts or fragments) were recovered from several sectors: Sherwood (14 passengers and 4 local residents), Rosebank (61 passengers), Beechgrove/Golf Course (43 passengers), Tundergrath (106 passengers), and Halldykes (27 passengers). Although the Pan Am crash is usually referred to as "Lockerbie," the fact is that more victims were found in Tundergrath than anywhere else. It was also in Tundergrath, four miles from the town, that the cockpit (the symbolic photo of the crash) fell (Gilchrist, 1992).

When there is warning of disaster, police personnel should be sent to safe locations near areas targeted for further deployment. Immediately upon declaration that the area is safe to enter, police can then be rapidly dispatched to their pre-determined posts. This allows perceived protection against looting and furthers the image of the police caring for the area of their jurisdiction (Greenberg *et al.*, 1990).

When there is no warning to a disaster, as in the case of many transportation disasters, responding personnel should report to a rear staging area, unless otherwise instructed. This prevents undue congestion at the scene, and in the long run allows for more effective work to be done. At the Avianca crash in Long Island too many fire-fighters and ambulance personnel reported directly to the crash site (Meade, 1990),[4] creating gridlock.

[4] Personal conversation with Samuel J. Rozzi, Commissioner, Nassau County Police Department, Mineola, New York (USA).

Given modern technology, it is now possible to input maps into computers, so that primary travel routes and staging areas can be identified immediately. Many areas are covered by commercially available programs.

Unusual field conditions often complicate police deployment and operations. Following the Edmonton tornado, a liquid nitrogen gas (LNG) tank was hurled more than a kilometer and a half in the air and landed on a road between two power lines. A major traffic jam ensued, and many police cars were neutralized. In all there were 14 geographically separated Hazmat incidents, but there were only three response units available (under usual circumstances a sufficient deployment) (Scanlon, 1992).

Responding forces at a disaster are typically stationed in concentric circles denoting different levels of response, with entry/exit checkpoints allowing passage between circles. For example, the innermost circle contains forces involved in immediate response. This would be police evidence collection, medical response, and fire. The next circle moving outward from the incident would be staging for those who have to enter at a later stage, with different functions or as relief/backup for first-stage responders. If the disaster requires a large number of responders, ambulances in the second circle might be limited to a certain number, so as not to cause congestion. More ambulances would be kept in the third circle (not necessarily concentric and possibly located slightly further away from the incident). An advantage of this system is to limit incident conversion, including that of uniformed responders.

One final word on the subject. Police officers are human. When there is possible danger to their families, they assume the traditional roles of parents. They are concerned about the safety of their families. This reaction is absolutely normal, however it is rarely mentioned in professional literature. (This author, for example, did not report for one disaster response until he verified that his son was safe.) This has the effect of delaying the police response.

On 19 May 1950 a munitions storage area exploded in South Amboy, New Jersey (USA). One responding policeman, later to become chief of police in the town, admitted openly that after hearing the explosion he spent time verifying that his family was unharmed; only after learning that they were safe did he report for duty.

CRIMINAL UNTIL PROVEN OTHERWISE

In many disasters there is a criminal act involved. Investigation after several earthquakes has shown that building collapse was caused in good measure by construction with substandard materials in violation of building codes.

Sometimes a disaster is caused by criminal negligence. And sometimes the disaster is caused by straightforward criminal intent. These acts, requiring police interaction with regulatory agencies, can be prosecuted in many localities, and again evidence is required.

Terrorist actions are criminal. Very simply put – when a terrorist's bomb explodes a plane in mid-air, the crime is murder. Clearly, murder must be investigated and prosecuted. When a terrorist is suicidal, the criminal investigation is not over with his death. In many cases the bomber does not work alone. There can be extensive support mechanisms behind him, ranging from safe house management and site survey to the obtaining of explosives and construction of the bomb. All of the people involved are accomplices, and they are the objects of police investigation.

To enable investigation and possible eventual prosecution, every disaster site must be treated as a scene of crime until proven otherwise. The area must be documented, and possible evidence must be collected. In many cases there is a psychological urge (if not political mandate) to restore normalcy as quickly as possible. This can be removal of a bus after a bombing or bulldozing rubble after an earthquake.[5] There is definitely a requirement that life return to normal, however sometimes extra time is needed. Premature clean-up should be undertaken only with the explicit knowledge that it will potentially destroy evidence. Admittedly, a disaster site is no ordinary scene of crime. If the definition of disaster is a disruption in daily life, then that disruption also affects police response.

[5] This was the case following the August 1999 earthquake in Turkey. In many areas rubble was removed before it could be determined if there had been building code violations.

A large disaster poses its own problems. Unlike a usual murder scene where relatively few people are present, a disaster site generates significant "traffic" (if only medical, fire and other emergency services) that can destroy significant evidence. This is complicated by the widened scope of response – people who do not usually attend scenes of crime and are less sensitive to evidence collection.

In many cases the cause of an incident is unknown until examination of the physical evidence and personal testimony collected shows cause. Initial reaction in Lockerbie, for example, was that "metal fatigue" caused the crash of Pan Am 103. Evidence collection and examination then showed that the true cause was the explosion of a bomb in the baggage hold.

OPERATIONS AT THE SCENE

Setting up response operations at a disaster site is a complex procedure that involves numerous agencies, each of which has its own response protocol. There are, however, certain general rules that apply to all responders.

The police almost always plays a central role in disasters. In most jurisdictions they bear nominal overall responsibility at least in the initial stages of operations. It is the job of the first arriving policeman to survey the situation and issue an appropriate call for help. This is an obvious step so that appropriate responding forces can be dispatched to the scene, but sometimes it is clouded by emotionalism (Coglianese and Howard, 1998). There is a major temptation to start administering first aid. Even so, surveying the situation and reporting should take precedence so that professional responders can arrive quickly. The following examples illustrate the problem:

- *Professional response.* The first responder on the scene of the Amtrak crash in Maryland (USA) described her choice of surveying the scene and reporting, rather than administering first aid, as the hardest decision of her life.
- *Emotional response.* In the Air Florida crash an emotional paramedic asked for 30 response units in addition to those already responding. This was obviously requested without site survey, since it was far in excess of real needs. If the request had been fulfilled, it would have unnecessarily tied up medical response for the entire metropolitan region.
- *Under-estimation.* The opposite mistake was made in the Metro train derailment that occurred within a half-hour of the plane crash. The initial request was for one ambulance; it soon became evident that although three died (two at the scene, one in hospital), another 22 were brought to four different area hospitals (DeAtley, 1982).

Communications is one of the greatest problems not only at a disaster, but in any large operation. Communications must be understood in its greatest sense – not only the ability to convey verbal messages, but also the capacity to convey non-verbal meaning. Uniforms and identification vests are a part of this non-verbal communication. An example is Hillsborough, where crowds had trouble distinguishing between police and ambulance personnel due to the similarity of their uniforms (Walsh, 1989). This similarity misses one of the major purposes of a uniform – to quickly convey the affiliation and task of the person wearing it.

Coordination between police, fire, and ambulance responders at a disaster site tends to be good, since these groups work together on an almost daily basis. That interaction builds rapport and an understanding of each group's needs. However, the further away from the scene (in terms of time or distance), the more problematic the coordination. Off-site responders (such as those in hospitals) have more problems with coordination than those working on-site.

Introduction of "disaster only" responders (e.g., aviation investigation) requires developing and drilling a new communications model. It also means resolving latent organizational concerns such as competition for prestige and limelight. This competition is more than organizational ego. It can have practical implications for future budgeting.

THE SITE(S)

Many novices to disaster response envision a stereotype single-site situation. That is not necessarily the case. Sometimes a disaster has multiple sites: in an earthquake, there is usually a broad area that is affected, with sub-sites of varying severity within that area. The area is usually more or less contiguous. Aviation disasters can present not only multiple sites, but even locations distant from each other.

[6] Former name of today's JFK (Kennedy) Airport.

Perhaps an extreme example is the 16 December 1960 mid-air collision of United Airlines Flight 826 (DC-8 from Chicago O'Hare to New York Idlewild)[6] with TWA Flight 266 (Super Constellation) over Staten Island, New York. The TWA aircraft fell in Miller Field, Staten Island; the United flight came down in the densely populated neighborhood of Park Slope, Brooklyn (Fodor, 1999). The result was two disaster sites separated by several kilometers and the bay waters of New York Harbor. The distance between the sites dictated use of separate hospitals and mortuaries. In essence it was as though there were two different disasters.

Crowds rushed to the site. Curiously, television reporters were allowed to enter the site, yet newspaper reporters were excluded. One "eyewitness" gave vivid accounts of destruction and tragedy; 39 years later he candidly admitted that all was fabricated. He had seen absolutely nothing (Fodor, 1999).

At the scene it is critical that police and other responders work according to a preplanned checklist of actions and jurisdictional assignments. This does not negate the necessity for changes necessitated by site conditions. The intent is "to limit uncoordinated and unnecessary activities" (Nassau County, 1995).

It is also recommended that each responding agency assign at least one or two emergency responders at a disaster site with the specific task of documenting the incident, so that critical evaluations can be made at a later stage and translated into improved working procedures. Although each participating unit should do this for its own internal purposes, the police are usually in the best position to deal with overall incident-wide documentation, due to freedom of movement and the availability of cameras. Video footage can give an overall impression of what happened, but there is no replacement for traditional still photography to record details of non-moving objects. A scale should appear in all photographs that highlight something specific (as opposed to general overviews). Sketches may also be useful.

As explained, the police must consider the disaster site as a possible scene of crime, and undertake the following:

- interrogate witnesses to establish who is responsible;
- determine what happened;
- impound evidence for possible prosecution.

At times it is necessary to discontinue utility services to render a disaster site safe. In such cases it must be ascertained that gas or electricity should not be resumed until specific authorization, properly coordinated with responding agencies, is given.

POLICE COMMAND

It is important to have a fixed-point command center, so that policemen and other responders find the police office relatively easy to locate. Yet, it is also necessary for commanders to tour the disaster scene to get first-hand impressions upon which operational decisions can be based.

EVIDENCE COLLECTION

One method of evidence collection advocated in professional literature is to overlay a grid on the disaster site, by stretching horizontal and vertical tape and labeling boxes from "A" in one direction (columns) and from "1" in the other direction (rows). This can be effective in an isolated area, however particularly in cases of urban terrorism where there is a rush to return to normalcy, the method can be too time-consuming to be practical.

The amount of evidence to be retrieved must not be under-estimated. This has significant implications for manpower deployment at the scene, the supply of evidence collection bags,[7] and office support procedures before the bags are sent to the various units dealing with them.

[7] Consideration must be given to both the quantity of evidence collection bags and the material from which they are made. Papers are collected in one type of bag. Human remains are collected in a very different bag. The size of bags is another consideration.

On 7 December 1989 a flight between Los Angeles and San Francisco crashed leaving 43 dead. Debris was strewn over 20 acres, and in all 60 tons of evidence were eventually collected. This included more than 840 kilograms of human remains. Work was so exacting that the copper jacket of a bullet was also found.

Sometimes special equipment is needed to remove evidence. This is true in most transportation disasters.

The net weight of the Boeing 747 that crashed at Lockerbie was 318 tons, plus load; 80 per cent was eventually recovered with some pieces found as far as 100 kilometers away from the main crash site. Construction equipment was needed to remove some of the airplane parts.

TRAFFIC CONTROL

The 4 January 1987 Amtrak train crash in Maryland (USA) is a good example of a case in which, according to at least one report, the police lost sight of one of their basic missions.

The first person on site was, by sheer happenstance, a young emergency medical worker. She filed an incident report, describing the situation and what response forces were needed. So far, so good. The response started properly. Police arrived within two minutes, but rather than setting up a traffic pattern for arriving emergency vehicles, policemen started administering first aid.[8] Again, establishing a traffic pattern to allow EMS personnel to enter the area is a step that benefits the majority of victims, rather than the few who might receive first aid. Traffic control should have been established.

In all fairness the last major accident in the local jurisdiction was an air crash 21 years earlier – beyond the normal limits of institutional memory, given personnel turnover. Thus, the experience accrued from the earlier incident was lost upon responders. Part of this experience was how to deal with insurance companies, federal transportation regulatory agencies, and commercial corporations – all groups with which the police do not usually interact.

There are endless examples of incidents in which traffic control is not instituted at an early stage. At the Kansas City (Missouri, USA) Hyatt Regency collapse on 17 July 1981 (Waeckerle, 1983) the parked police cars of officers responding to the disaster had to be towed away to allow access for other emergency vehicles.[9] At the Avianca crash site in Long Island some emergency equipment had to be carried by hand for more than one mile due to traffic congestion (Meade, 1990).

Air traffic must also be taken into consideration, as helicopters in particular respond to disasters. This can overburden normal air traffic control.

[8] The police forgot another basic mission standard in any operation. Only on the day following the accident (after one of the trains had been towed away), did the police realize that there might have been criminal negligence involved. Eventually, the engineer and brakeman were convicted of 16 counts of manslaughter.

[9] A practical piece of equipment as part of a disaster plan is a truck to remove parked cars, in this case even those parked legally but impeding the flow of emergency vehicles.

On 17 July 1981 as some 2000 people gathered to watch a dance contest, the second- and fourth-level walkways of the Kansas City (Missouri, USA) Hyatt Regency Hotel collapsed onto the ground level; 114 people died and at least another 200 were injured.

SCENE CONTROL

A second immediate job for the police at a disaster site is to close off the site. This is done for several reasons.

- Limiting the number of people in the area reduces confusion and allows greater mobility of those who need to be in the area.
- Evidence must be guarded.
- Personal property is often abandoned. (The owner, for example, may have been taken to hospital.) This property must be safeguarded and eventually returned.

Closing a site is possible in a limited area disaster; it is totally out of the question in an area-wide disaster such as in a mid-air crash in which materials inevitably fall over a large area.

There are numerous medical hazards connected with a disaster site. In an extreme example, a spectator died of an apparent heart attack following an airplane crash (American Airlines Convair) in Elizabeth, New Jersey (USA) on 22 January 1952; he was pronounced dead on arrival at hospital. There are also hazards to the crowds who converge around the scene to catch a glimpse of the response. If nothing else, this is a reminder that certain basic medical services should be kept on standby for contingencies once disaster life-saving operations have been completed.

Not every person at a disaster site is who or what he purports to be. On 24 June 1975 two persons, one claiming to be "an emergency worker" (with a medical bag) and the other "a policeman," were arrested for impersonation (Grayson, 1988, p. 126) following an Eastern Airlines crash at New York JFK Airport. At the Skywalk collapse in the Kansas City Hyatt Regency Hotel on 18

Eastern Airlines Flight 66 (Boeing 727 aircraft) crashed into the approach lights to runway 22L at New York-Kennedy Airport during a thunderstorm. The accident killed 113 of the 124 persons aboard.

July 1981 a reporter was caught wearing a Red Cross vest; his objective was to obtain names of victims. In the Egyptair crash a reporter was arrested after pretending to be a bereaved family member.

Sometimes the problem is not impersonation, but rather abuse of position. After one bus disaster in Israel an employee of a senior government official used the name of her boss in an attempt to gain incident information for personal purposes.

Once volunteers and passers-by finish their tasks at the site and regular forces take over, volunteers should, according to pure theory, leave the area. This process poses a critical question for professional responders. They are faced with the problem of removing from the scene the very same volunteers who were of potentially critical importance before the professionals arrived.

CROWDS

Disaster response involves two types of amassing crowds. The first type is the crowd that gathers, forming a potential for disaster. The second is the crowd that forms after a disaster. Crowd disasters tend not to occur on transportation vehicles, but they are a distinct possibility in transportation stations and terminals.

Losing control of a crowd can create a dangerous situation, particularly when the crowd surges or stampedes. An extreme example happened on 3 March 1943, when sirens were sounded in London after two nights of Nazi bombing. In 10 minutes alone some 1500 people fled into the Bethnal Green underground station. Experimental anti-aircraft batteries opened fire, more Londoners surged into the shelter, a woman holding a baby tripped, and 173 were killed in the ensuing crush. On 3 July 1990 1426 pilgrims on the Haj in Mecca were killed in a crowd stampede that started in a pedestrian tunnel.

Despite these examples, crowds are a social phenomenon whose behavior and dynamics can usually be predicted. As a result, there are basic guidelines in crowd control to minimize the possibility of disaster, such as watching points where different paths converge and avoiding unsupervised reversal of direction in crowd flow (e.g. the reversal from event entry to event exit must be a supervised transition done during the event at a point with virtually no traffic).[10] Crowds should also not be considered as a single entity; very often there are subgroups within a crowd with their own identifiable structure, conduct, and sometimes even leaders. Very rarely do these "leaders" seek to send a crowd out of control. More often there is an effort to manipulate a crowd into voicing slogans (after terrorist-caused disasters often decrying poor security) or into specific actions. It is the leaderless crowd, not willing to listen to police or other instruction, that poses the greatest threat.

[10] In a transportation example, in times of very heavy passenger traffic, exit from a rail car in station must be separated from entry.

Crowds gather either as part of an existing event (such as a soccer game), or in an action where the event is that gathering, as in a political demonstration or rally (US Army, 1968). In the former case it is much easier to control the crowd safely. In the latter case, even in a demonstration "friendly" to law enforcement, organizers often want a certain controlled escalation of crowd-gathering.

POST-DISASTER CONVERGENCE

The gathering of crowds around a disaster site is common, as young and old try to get a glimpse of what happened. Sometimes they are able to see. Sometimes they just want to be seen. Sometimes it is just the excitement of the experience that motivates them. Some spectators misbehave. Some spectators behave without problem. Some spectators even volunteer their services. The only constant is that there always seem to be crowds at disaster sites – certainly those in or contiguous to populated areas – and when there are crowds, police must deal with them, if just to allow for free passage of disaster responders.

> Without the convenience of modern transportation, crowds gathered in Asbury Park, New Jersey (USA) in 1934 to see the consequences of the *Moro Castle* ship fire. On the first Sunday after the fire an estimated 150,000 to 300,000 people gathered. A week later on the following Sunday estimates were as high as 500,000 people converging on Asbury Park (Lindbloom, 1950a).

> On 31 August 1986 there was a mid-air collision over Los Cerritos, California (USA). An Aeromexico DC-9 flight approaching Los Angeles International Airport collided with a Piper PA-28, leaving 82 people dead.

John McGuire, a California fire engineer, gave a very picturesque description of crowds at Los Cerritos following the air crash: "People were just like animals. If anybody's ever been on a big brush fire or mass casualty or anything like this, they know that people flock to it like ants on a cake, and they do more harm than good. All they do is get in our way" (Brooks, 1986).

Calling the people in crowds "ghouls and fetishists" (Taylor and Frazer, 1980) makes no constructive contribution. Understanding the crowd as an expression of *convergence* is not merely an academic exercise; it is a determining factor in deciding how to handle that crowd.

Crowds and convergence are not restricted to pedestrians. Michaels (1990) reports "helicopter convergence," constituting an airborne crowd following the Loma Prieta earthquake. Helicopter pilots used their aircraft to get a "better view."[11]

[11] This constituted both a safety hazard and complications for response forces.

Convergence might be bothersome, but it is not threatening by definition. If people want to watch action, only objective reasons should be used to prevent them congregating in a defined area that does not interfere with the response. As in many cases, trying to prevent a natural psychological reaction (such as convergence) will only inflame the situation without achieving any tangible gain. Convergence that verges on disorder clearly must be handled with a much less *laissez faire* attitude.

Both crowds and police can be overcome by emotion or by a misdirected decision of what is expected from them. When this happens, the result is usually confrontation.

LOOTING

There has been considerable discussion as to whether looting exists after a disaster (Granot 1993). For the purpose of discussion the following definitions should be used: (a) *theft* is the clandestine taking of property, whereas (b) *looting* is the unrestrained and open taking of items belonging to others in an atmosphere of the breakdown of law and order (when deviant behavior becomes the norm).[12]

[12] An extension of the military model of plundering.

Following the San Francisco earthquake of 1906 Mayor Schmitz issued orders for "quick justice" to prevent looting. Morris (1906/1986) describes instances of immediate hanging following the earthquake, which has sustained the folklore of a strong hand to maintain law and order. The reality is that the examples were of theft of valuables from bodies and not of wanton looting.

On 4 October 1918 there was a major explosion of munitions (in three stages, at 1930, 0525, and 1025) at the T.A. Gillipsie Loading Company in Morgan, New Jersey (USA). One estimate was 87 persons killed. Some reports highlighted extensive looting, but this seems quite questionable in view of another report that all area residents (50,000 people) fled, and only fire-fighters remained. The fact that the main munitions magazine was in danger of exploding constituted an imminent danger, again casting doubt on a scenario of people staying behind and trying to steal (Lindbloom, 1950).

Two Long Island Railroad trains crashed on 22 November 1950 near 126th Street and Hillside Avenue in the Richmond Hills section of Queens, New York (USA). Although there were reports of looting, it is more probable that it was theft, since there was no major breakdown of police control.

On 5 December 1978 a blast occurred in a refinery in Linden, New Jersey (USA). The *Star Ledger*, a New Jersey newspaper, reported looting as far away as Bayway, where store windows had been broken. There is a theoretical question whether this was theft or looting; however from a law enforcement perspective, the exact definition has secondary importance. In either case, police deployment was necessary. Perhaps the true determination of theft or looting is in the nature of the police deployment.

On 24 May 2001 the floor of the Versailles Hall, a catering establishment in Jerusalem (Israel), collapsed during a wedding reception. Twenty-three people died and hundreds were injured. During the police investigation it became clear that monetary gifts and perhaps purses had been stolen. This is another example of theft (done clandestinely) and not looting.

To keep the question in its proper proportion, following the Loma Prieta earthquake (technically felt over a 400,000 square mile area) only two of the over one hundred law enforcement agencies with jurisdiction reported property crimes (Guerin, 1990).

The definition of looting is not without problem. It is common to talk about looting, but seldom is the term defined. On 29 December 1972 Eastern Airlines Flight 401 crashed in the Florida (USA) Everglades. "Looters were seen taking jewelry and wallets from bodies around the disaster area." A Coast Guard worker saw someone taking watches "and things" from victims (Grayson, 1988, p. 93). Although this is described as looting, the behavior of the Coast Guard worker raises a question. He made the conscious decision that saving lives was an overriding priority. In other words, there was no atmosphere permissive to theft; there was simply a reordering of working priorities.

Following Lockerbie there were several arrests for theft of crash "souvenirs," including subsequent convictions in court. It was also reported[13] that there were bereaved family members who took unauthorized "mementos" home as a memory of the crash (in harsh legal terminology, theft of evidence in a criminal investigation and subsequent prosecution, but sympathetically tolerated though not condoned under the circumstances).

[13] Private conversations.

The Beverly Hills Supper Club in Southgate, Kentucky (USA) caught fire on 28 May 1977 (for the third time in seven years), leaving 165[14] people dead and at least 30 with injury needing hospitalization. The club had been built in 1937 and was certified for an occupancy of 536. Reports vary however the number of occupants at the time of the fire was at least 1350. Unusual heat was felt at 2015. Thirty-five minutes later there was smoke. At 2102 flames swept into one of the dining rooms.

The fire and ensuing deaths was a system failure. Causes included: overcrowding, substandard wiring, flammable building materials, a shortage of exits (partially due to locked doors).

[14] Two unborn fetuses were also killed, raising the question if they should be included in the death count (included in some sources, in others not).

In the first four days following the Beverly Hills Supper Club Fire there were rampant rumors that private possessions had been stolen from the dead. In fact, however, there had been no looting. Two people were arrested for theft, however they were soon released, and charges were dropped. (Another baseless rumor was that in the panic to escape, some of the victims had been clawed and clobbered by others trying to escape; the truth was that there was no evidence of panic, nor was there any indication of physical assault.)

Claims of looting at the Supper Club also raised a paradox. On the one hand surviving family members believed the rumor that wallets had been stolen from bodies. On the other hand those same supposedly missing wallets were a key element in the identification of the male bodies. This contradiction should be considered as an expression of people's fear, their attitude towards property, and an expression of the disaster myth of looting.

Another disaster myth is that people *panic*. Sime (1980) has described panic as a *social myth*. This is a very rare phenomenon, and found only in cases of fire that closes escape routes.[15] More often people are directed by calculated self-interest, motivated by survival and selfishness that disregards others. An early example of supposed panic is the 30 June 1900 fire at the dock in Hoboken, New Jersey (USA) that left some 400 people dead. There were reports that panic broke out when winds changed direction and engulfed fleeing crowds in flames.

[15] Not a universally held opinion. Some analysts view reaction to lack of escape as a frustration caused by lack of options rather than panic. According to that theory, if resultant behavior is "irrational" according to normative models, that does not mean that the behavior can be described as panic.

Looting is also cultural. Following the 1983 crash of Gulf Air Flight 771 from Karachi to the Gulf area, in the desert between Dubai and Abu Dhabi (United Arab Emirates), there was extensive looting. The phenomenon of looting was also true of the Thai International crash in Nepal and the Lauda crash in Thailand.

On 23 September 1983 a Gulf Air flight (Boeing 737-200) crashed upon landing approach to Abu Dhabi Airport. A bomb exploded in the plane's baggage compartment. All six crewmembers and 105 of the 111 passengers were killed.

According to Quarantelli and Dynes (1970), if there is looting (in Western society), it is done by people coming from outside the disaster area. Hurricane Camille of August 1969 is often cited as an example, with reports of looting by National Guardsmen brought into the area.

If looting does not exist, profiteering certainly does. There are numerous examples of charging outrageous prices for commodities when they become scarce after a disaster. There are, in fact, cases in which the method of distributing relief supplies only aggravates the situation.

There are numerous examples of artificial price increases. Following the onset of the "Al Aqsa Intifada" in Israel and the Palestinian Authority in September 2000, hotel prices fell in Israel due to the lack of tourists; prices in Gaza, however, rose by 25 to 30 per cent as quality hotel space became scarce due to the influx of reporters.

INVESTIGATIONS

Interrogation of witnesses is an aspect of evidence collection and part of determining what has occurred. Determination of what happened in the initial stages of a disaster is by no means an easy task. In many cases eyewitness testimony is unavailable – because no one was present, because those present did not see or understand events, or because those present perished in the disaster. In such cases forensic examinations can sometimes unravel mysteries.[16]

In the short term deciding what happened has implications for terminating danger. In the long run, that determination can affect accountability, prosecution, and possible safety measures designed to insure that the disaster does not

[16] The investigatory role of forensic science should not be confused with disaster victim identification (DVI), discussed elsewhere.

repeat itself. Sometimes the determination takes months if not more of work by commissions with attached professional experts. In any case, it is usually the police that undertake initial inquiries.

Most of the standard work conducted by an investigator at a standard crime scene (Fisher, 1993) is relevant at a disaster. He must determine what happened and who was responsible for certain actions. There are, however, certain basic differences.

At the crime scene:

- There is relative quiet in which to work.
- Few people are present.
- There are usually few witnesses to the crime.

At the disaster site:

- Tumult and confusion are typical, particularly in early stages.
- Many people are present, and crowds usually form.
- There can be many witnesses. They can be hard to find amongst the gathering crowds. Much of the testimony is typically repetitive and very generalized.

Despite the activities and pressures of a disaster site, testimony must still be written clearly (and not in shorthand) and recorded with required signatures, so it can be later used in court.

PHASING DOWN OPERATIONS

In many disaster responses there are three basic stages of operational intensity:

1. Activity hits a very quick *peak* as forces quickly converge on the scene. This can be in the moments before an airplane attempts an emergency landing, or after a train has derailed. Basically, responders hurry to the scene and administer whatever assistance possible.

   Sometimes activity can shift from one function to another. Different functions can have different peaks at a disaster. Medical response has one peak. Body removal has a different peak.
2. At some point, however, many of the responders are no longer needed. The operation must be *phased down*. When operational circumstances permit, the general personnel staffing of the response, such as policemen assigned to insure public order, can be reduced.
3. The *end* of a response is often evident. Ambulances with their assigned

> crews are no longer needed. In police terms, the wreckage has been removed, those responsible have been identified, the crowd has dispersed. Releasing ambulance teams is a relatively easy decision. Reducing the ranks of police responders is a much more difficult response.

Any disaster response gathers a momentum of its own. It is most difficult to assess the situation, take the initiative, and phase down operations, as requirements lessen in a disaster response. Phasing down is also not a one-time procedure. It can be a gradual process over an extended period of time.

> The *New Carissa*, a 600-foot bulk freighter, ran aground one mile off the North Point Jetty at Coos Bay, Oregon (USA) on 4 February 1999. The ship carried 23 people on board; no one was injured. An oil spill ensued, and a Unified Command was established to handle the response.
>
> Cleanup, diving and salvage operations continued for months. On 11 July 1999 the US Coast Guard recommended phasing down, by closing the full-time Unified Command. The proposal was met with opposition and was referred to the Oregon Department of Environmental Quality for decision, rather than remaining in professional operational circles.

CONCLUSION

Police play a dominant role at any disaster site, as they handle numerous tasks, ranging from traffic, to security, to investigations. In many jurisdictions it should also be remembered that various police departments are responsible for different activities.

The police are, however, only one of many "players" present. Other functions, such as medical and fire also have responsibilities.

CHAPTER 19

FIRE

INTRODUCTION

Fire is often a significant part of a transportation disaster. In general terms, post-crash aircraft fires are more lethal than crash impact. Fires have destroyed ships at sea. Fires have also accompanied train, plane, truck and car accidents. Thus, consideration must be given to this aspect of transportation disaster response.

BACKGROUND

People have a certain fascination about fire. On the one hand, sitting next to the fireplace is often used as an example not only as a sign of physical warmth, but also as an image of the social warmth of a household. Once one gets too close or that fire gets out of control, that scene of serenity can turn into a story of indescribable horror bringing death and destruction.

Throughout the centuries man has tried to harness fire, first as a source of heat and light, then as a component in power and energy. One major triumph was the discovery in 1827 by John Walker (Stockton, England) of "lucifers" – the friction match (a mixture of potassium chlorate, antimony sulfide and gum).[1] But with increased use of fire came increased danger. With increased facility to produce fire came an increase in misuse – arson.[2]

Fires are a common element in disasters. Sometimes fire is the primary cause of the disaster. In other cases there is another primary cause (e.g. air crash), and fire is a secondary phenomenon. This distinction has implications for theoretical analysis and future mitigation, but in response it makes little difference if fire is primary or secondary. It must be extinguished, and potential victims must be distanced from it.

In modern society it is common to stress technology. Science has introduced a long list of inventions that have become important facets of everyday life. These range from the microwave in the kitchen to the cellular telephone in someone's pocket. They range from laptop computers to jet aircraft. People think in terms of technology. When it comes to disasters, the tendency again is to think in terms of technology – to look for a technical cause to determine why

[1] The phosphorus match was invented by Charles Sauria (France) in France in 1831 and patented in 1836.

[2] In the nineteenth century two other technological developments had an effect on arson – the production of paper at a very inexpensive price and the ready availability of petroleum products.

an aircraft caught fire or why the Hazmat cargo of a truck exploded. Very often this search for technical reasons ignores the human factor. Many disasters, if not caused by human error, are at least aggravated by poor human decisions.

A fire can be *intentional* (as in the case of *arson*) or *unintentional* (e.g. an electrical short). There is, however, middle ground. Although the Beverly Hills Supper Club fire was caused by electrical problems, there is room to argue that trying to deal with the fire in-house for 20 minutes before reporting it to the fire department was problematic behavior, stated in very modest terms. While arson was certainly not involved, this behavior has to be taken into account in considering responsibility for spread of the fire and resultant death and injury. The Board of Inquiry following the disaster was harsh in its judgements – behavior at the facility was incorrect and certainly contributed to the severity of the situation – however, no criminal charges were brought. Transportation disasters are no different. They, too, are very much influenced by the human factor.

In an aviation scenario, for example, what should a pilot do when he detects smoke or possible smoke? Should he continue flying as he tries to determine the source of the smoke (in some instances harmless), or should he immediately initiate emergency landing procedures (knowing that the source might be trivial)? This was a dilemma faced by the pilot of the ill-fated Swissair 111 flight as documented in the flight recorder.

COMPONENTS OF FIRE

A traditional description of fire includes a triangle reaction: heat, oxygen and fuel. A typical fire will have these three components.

The most common method of extinguishing fires is to reduce oxygen levels. An exception is gas leak, in which case the prevalent method is to cut off fuel supply.

ARSON

Burning[3] caused by arson (burning as a result of a willful and intentional action) is a clear criminal act. There is a wide range of motivation for arson. In transportation cases the most common motivations are financial gains and terrorist goals.

When arson is mentioned, many people immediately think in terms of buildings. Yet, according to the United States Department of Justice, in a typical year vehicles account for between 15 and 25 per cent of all arson cases in the United States. The motive is usually insurance payment. These vehicular arson cases include single vehicle incidents (often in desolate areas with the driver miraculously leaving the vehicle before it is consumed by fire).

[3] In most jurisdictions this includes "fires" that do not have open flames, such as smoldering wires. The presence of burning is relatively easy to show. Intent/willful action is more difficult to prove. Even when proven, that action must then be associated with a specific person before there can be a prosecution.

Investigation of Vehicular Arson

- Determine if quality tires have been replaced.
- Has the gas cap blown off as happens after a fuel explosion?
- Was gasoline released from the tank to pool under the vehicle?
- Did the fire start in more than one place?
- Was there a flashback caused by pouring an accelerant?
- Are any parts of the car missing or removed? (radio, tools, engine parts, etc.)

(O'Connor, 1987)

Sometimes putting out arson fires is not simple. An example of mass disaster due to arson can be seen in the two petrochemical plant fires that were ignited near Colombo, Sri Lanka in the early hours of 20 October 1995. Gun battles between terrorists and security forces increased the dangers posed by the fires and took the toll of at least 25 people. In another example, in September 1970 the PLO hijacked Western aircraft that were forced to land at Dawson's Field, Jordan. Hostages were eventually released, and the aircraft were set on fire. Government forces could not gain access to the area to extinguish the flames.

Arson is usually considered a clandestine activity done by someone who wants to hide the origins of a fire. The terrorist phenomenon of throwing Molotov cocktails – fire bombs[4] in simple language – contradicts this image of clandestine action, although those throwing the incendiaries do make every effort not to be caught. These fire bombs, often bottles filled with flammable materials ready to ignite, are thrown with the object of inflicting damage.[5] Vehicles are often the target of Molotov cocktails, the perpetrator hoping to ignite the fuel system.

[4] Molotov cocktails (originally a Second World War "weapon") are usually bottles filled with gasoline with a fuel-soaked rag in the neck. Another version is a mixture of one part gasoline and one part motor oil.

[5] A proper response to a fire bomb incident includes forensic analysis of the bottle contents for subsequent courtroom presentation.

WHERE FIRE BREAKS OUT

Building fires can be in: (a) licensed facilities, (b) unlicensed establishments (such as the fire in the Happy Hour Supper Club in the Bronx), or (c) quarters licensed for a purpose other than the one at the time of fire (as in a clothing factory in Hebron, Palestinian Authority where the area was licensed for cold food storage). Licensing is a part of mitigation. As the latter example shows, however, mere issuance of an operations license is meaningless unless there is inspection and a requirement for safety measures appropriate to real use.

Considering fire risks before licensing exists in transportation as well. In Israel, for example, vehicles authorized to carry 11 or more passengers must have a fire extinguisher. This is in addition to the numerous regulations of manufacture that protect vehicle fuel supplies and prohibit extremely flammable upholstery. Parallel fire mitigation measures are also taken in aircraft design before use authorizations can be issued.

Sometimes a fire is relatively small, but the basic cause of injury and death is to be found elsewhere – such as the closing of emergency exits (inspection of which is one of the parameters of licensing). A parallel to this is an aircraft, in which emergency exits are relatively few with, at best, two aisles in which passengers can move.

This phenomenon of closed exits occurred in the Iroquois Theater in Chicago, Illinois (December, 1902), the Cocoanut Grove Nightclub in Boston, Massachusetts (November, 1942), and the Bradford Stadium in England (May 1985). Although these incidents span almost a century, they show than an obvious safety lesson has still to be learnt. Both the Iroquois Theater and the Cocoanut Grove Nightclub were decorated with highly flammable fabrics – another lesson that went unlearnt (Taft and Reynolds, 1994).

On 28 November 1942 a fire broke out in Boston's Cocoanut Grove nightclub, killing 492 people. The facility had an official capacity of 600, but on the night of the fire it was reportedly occupied by more than 1000 persons. After flames broke out, the crowd ran for the exits, only to find some of them sealed. Some observers reported the flight that ensued as panic, with many deaths due to the stampede and crushing. Another primary cause of death was asphyxia from toxic gases being released.

This fire is considered an American landmark event in the legal fight to require adequate exits and emergency signs in restaurants and nightclubs.

Although the initial cause of the Cocoanut Grove fire can be considered technical (due to electrical wiring), Sime (1980) describes the ensuing "panic in flight" as being used as a scapegoat to avoid the issue of human failure.[6] *If* there was panic in trying to exit, it was because of the error not only in sealing exits, but also in their very planning. Proper thought had not been given to emergency flight. Piling up at known emergency exits was, in fact, a rational action with a result that could be logically predicted with any kind of professional foresight. Simply put, "panic" can be used to cover up faulty preplanning.

[6] In the same way, the term "natural disaster" is also used to cover up human failures. Natural hazards are known and must be taken into consideration in planning. When a natural event causes mass destruction and loss of life, it can be argued that what really has occurred is a failure in planning and building.

FIRE OPERATIONS

The stereotype of a fire is a conflagration put out quickly by arriving firefighters. An obvious exception is a forest fire, comfortably far from population centers, which does burn for considerably longer periods of time.

The Hagersville, Ontario (Canada) tire factory fire contradicts this myth. The fire, which consumed 14 million tires, lasted 17 days and involved the evacuation of area residents. It also caused extensive Ontario Provincial Police (OPP) deployment of forces, in addition to fire personnel. In this case basic safety was the function of Ontario Provincial Health and Safety, and not of the fire department. At one point Ontario provincial authorities halted work near the scene at Hagersville for 72 hours due to health considerations (Scanlon, 1992).

After report of a fire is received, it is confirmed (in cases of doubt about authenticity), and fire-fighting personnel/equipment are dispatched to the scene. In serious fires or cases in which there is a possibility of terrorism, the fire department senior officer coordinates with police and decides upon the degree of danger and level of fire department presence required. In addition to extinguishing any current fire, these usually involve identification of additional hazards and setting inner and outer circles for fire protection. There also has to be an approved plan for traffic flow, so that fire, medical and bomb disposal (when needed) personnel have unhampered access to the disaster site without blocking each other.

In cases of terrorist bombing it is essential that fire personnel work in close coordination with bomb disposal units in searching the area, and closing down electricity lines and gas mains. Any second explosion (even if an intentional controlled explosion by bomb disposal experts) runs the risk of causing secondary fire and/or explosion.

Transportation fires, however, tend to be in much more limited spaces; they inevitably pose an immediate threat to life, and effort must be made to extinguish them immediately.

There is one phenomenon that is common to many aviation fires – the exact source of the fire is unknown. In the 1998 Swissair crash, for example, the pilot smelled smoke prior to the crash, yet he had no way of knowing where its origin was in the plane. Even if he had known, he had no way of extinguishing the fire.

FIRE SUPPRESSION

Water is used to extinguish many usual fires. Certain Hazmat materials require special techniques. For this reason there are several foams commonly used to extinguish Hazmat fires.

- *Class A:* Hydrocarbon sufficants. Often used with compressed-air foam systems (CAFS).
- *Class B:* For flammable and combustible liquids.
- *Protein foam:* Resists burnback.
- *Fluoroprotein foam:* Fluorochemical sufacants added to protein foam.
- *Aqueous film-forming foam:* Most commonly used. AFFF spreads a thin aqueous film over fuels.
- *Film-forming fluoroprotein:* FFFP is based upon AFFF technology.
- *High-expansion foam:* Used underground or in bulk paper storage fires.

MEDICAL ASPECTS

In responding to fire it should be remembered that more people generally die from asphyxia due to smoke inhalation than from flames. In building fires, for example, the United States Fire Administration reported that for the period 1994–96, 22 per cent of victims died from asphyxia only, whereas only 2 per cent of victims died from burns only (US Fire Administration, 1999).[7] Benjamin/Clark Associates (1984) conducted a statistical study of fires in the United States in 1981. Their findings, representative of most other years, showed that only 5 per cent of fatalities were from open flames. Smoke inhalation was the major cause of death.

[7] Seventy-five per cent of the deceased persons suffered from both asphyxia and burns, however this statistic does not take into account the relative severity of each injury and the basic cause of death. Unconfirmed estimates, however, consider asphyxia to be the primary cause in virtually all of those deaths.

When carbon monoxide (CO_2) enters the body, it replaces regular hemoglobin with carboxyhemoglobin (COHb), thus causing death.

The worst example in the twentieth century is probably in the aftermath of the earthquake that struck the Tokyo–Yokohama (Japan) area on 1 September 1923. In all 99,331 deaths were counted; some 40,000 of these were people who crowded into an open area, where they were choked by smoke and by the dust it carried. Another classic example of this is the Happy Land Supper Club in Bronx, New York, where, on 25 March 1990, 87 people died after a fire started by arson – the throwing of an incendiary device following a lovers' squabble. All fatalities were from smoke inhalation.

Also in many airplane fires more deaths are caused by smoke inhalation than by the flames of a fire. It has been estimated that a significant number of people survive air crashes, only to die afterwards in the post-landing fires that break out (again more often from smoke inhalation than from burns). A key example is the crash of Air Canada in 1983.

On 2 June 1983 Air Canada Flight No. 797 en route from Dallas, Texas (USA) to Toronto, Ontario (Canada) made an emergency landing in Cincinnati, Ohio (USA) due to smoke and fire in an aft washroom. All 46 persons (41 passengers, 5 crew) aboard the plane survived the emergency landing. When fire broke out after impact, 23 people died within one or two minutes, primarily from smoke inhalation. That smoke contained toxic fumes and particles emitted from materials in the cabin. Ironically, opening cabin doors to allow rapid exit of passengers let fresh oxygen into the cabin and fueled the fire.

As noted earlier, not all victims succumb to smoke inhalation. People are also burned in various degrees. Burns fall into four categories:

1. First degree: reddening
2. Second degree: blistering
3. Third degree: scorching
4. Fourth degree: charring

From a response perspective, extinguishing many fires results in mud created by hose water. In one case of a bus bombing in Israel on a hot July day, rescue efforts had to be stopped every 15 minutes or so for the fire department to douse down the vehicle to prevent further fire. This created a serious amount of mud that complicated the response effort (Levinson and Shmeltzer, 1992). It meant a need to issue boots to responders. Mud, even that created by fire extinguishing, can also make medical evacuation stretchers on wheeled trolleys impractical.

In the same incident backup fire trucks stayed in the area even after the initial fire had been extinguished. This proved invaluable as a smoke grenade thrown by a technician to guide a medical evacuation helicopter rolled into the bushes, causing another fire.

DEFINING THE INCIDENT

Although one might take the simplistic attitude that anyone can see a fire, that is certainly not always the case. In some instances fires start within walls, and only when they are already burning strongly do they break through into the open. In the Kansas City Hyatt collapse, dust raised by the fallen skywalks resembled smoke (Gray and Knabe, 1981). At the crash of a Galaxy the fire on the scene at a Reno, Nevada (USA) trailer park was obvious; what was not obvious was that the source of the fire was a crashed aircraft.

MULTI-FIRE INCIDENT

Disaster does not always cause only one fire. There have been numerous cases of more than one fire burning at the same time.

Fires after the Aeromexico crash in Cerritos, California (USA) (31 August 1986) are an example of a multi-fire incident. The commercial DC-9 crashed into a Piper Archer light plane. Both aircraft came plummeting to the ground, where 16 homes were set on fire. From a fire-fighting perspective, it was necessary to handle many of the perimeter fires to gain safe access to the main crash site. Problems encountered included water pressure and that fact that different materials were burning (aviation fuel, home-use natural gas, etc.), requiring differing fire-fighting techniques (Brooks, 1986).

Due to the intensity of the fire, there was little for paramedic teams to do. People either survived by walking away with minor injury or perished in the flames. For that reason some of the paramedics helped with the fire fighting effort.

CONFUSION

Many fire scenes are accompanied by confusion. In a 1987 fire in a Bogor, Indonesia department store, 70 bags with human remains were evacuated to a hospital mortuary. Police were puzzled. All but ten of the store employees were found. How could there possibly be 70 bags with bodies? It seems that under pressure in the response, rescuers mistook mannequins for deceased persons! The death toll was officially set at ten (Fire International, 5/1996).

A similar reaction, this time aviation-related, happened after the Thai International crash near Kathmandu, Nepal. Only selected soldiers could make the hike to the mountain crash site; one older English responder perished trying to reach the site. These young and inexperienced soldiers collected numerous bags of "remains," which were brought back to the command center from the site. Most of the bags contained little more than mud and rocks. In one bag brought back to base there were human remains and a bone later determined by forensic experts to be that of a dog. It is commonly accepted that the femur (thigh bone) and tibia (leg bone) of humans and dogs are similar.

INCIDENT AFTERMATH

The area must be restored to "normal" after a fire. It is the responsibility of the fire department to ascertain that it is free from further fire hazards. Closing entrances, cleaning garbage, and removal of routine hazards (e.g. broken glass) are the responsibility of pre-designated authorities according to the decisions of each particular local jurisdiction.

CONCLUSION

Fire is a major problem in transportation. This includes not only extinguishing fire, but also the forensic examination of a scene to determine whether arson has been involved.

CHAPTER 20

FORENSIC SCIENCE

INTRODUCTION

In a most technical sense forensic science is the investigation, according to scientific principles, of possible evidence to be brought before the *forum* (Latin), or in modern parlance a court of law.

Forensic science can be used to: (a) help reconstruct what happened, (b) examine if the act that transpired fits the definition of a crime, (c) determine who was involved, (d) suggest to investigators directions of further investigation, and (e) present findings in courts of law or before other judicial fora.[1]

[1] It should be remembered that the forensic scientist presents his expert opinion. It is the judicial body that makes determinations.

DEFINITION OF CRIMES

In many cases a crime is involved in a disaster. This is blatant in cases of terrorism or mass murder.[2] It may also be true, however, in low-profile transportation accidents or even in natural disasters.

[2] As in the case of killing 62 tourists at Luxor, Egypt on 17 November 1997, which was both terrorism and mass murder.

- In a traffic accident there can be drunk driving (in most cases a criminal offense) or driving without a license.
- In a hurricane no one can blame the weather; in an earthquake one cannot prosecute "Mother Nature." But there have been numerous cases of engineering inspection uncovering sub-standard building construction after a disaster. (A frequently cited example is the Armenia earthquake.)
- There have been Hazmat incidents and vehicle/airplane accidents with implications of criminal negligence.

The police do not have a monopoly on forensic science. It is just that some of the best-known disciplines (e.g. ballistics, toolmarks, fingerprint development) are routinely handled by the police. Although forensic medicine is often organized outside the police, there is a strong working relationship between the police and the medical examiner due to routine casework. Ironically, many of the required forensic skills in disaster-related cases are not found in the forensic laboratories of either the police or the medical examiner (e.g. structural engineering).

The fact that not all legal violations in a disaster fall under police enforcement is yet another strong indication that a multi-agency approach is warranted.

In many localities private forensic experts are also available to conduct examinations, issue expert opinions, and testify in court.

SCENE INVESTIGATION

When individuals are involved in an incident, the golden rule is "innocent until proven guilty." This applies on the individual level, but not on the investigatory level. When police handle a disaster site or case involving death by other than natural causes,[3] the golden rule of the investigation is "assume that there is a criminal act involved until proven otherwise."

[3] Death by other than natural cause can be murder, suicide, or accident with various degrees of criminal liability involved.

Using this theory, it is critical to maintain site integrity. The area must be kept as intact as possible, assuring that a minimum of evidence is destroyed, and guaranteeing that no new material is added to the scene either intentionally or unwittingly.

Photography of the disaster as soon as possible is desirable, so there is a record of the disaster before fire, bomb, medical, and other first-priority responders alter the site.

EVIDENCE COLLECTION

One must maintain a documented and uninterrupted chain of evidence from the scene, to laboratory, to examination.

Just as in more standard criminal cases, in disasters all forensic examinations are dependent upon proper evidence collection at the scene. The better evidence collection is, the better chance there is of receiving quality results. In disasters this is, of course, more difficult given the number of people at most scenes and the life-saving activity going on in the area. In particular plane or train crashes create disaster sites that are much more extensive than in a homicide case.

A general rule at all disaster sites is that the area should be documented, i.e. photographed. Video can give an overall impression of the site, but a standard photograph is still best for recording fine details. Film or digital cameras also have the advantage of allowing enlargements of specific items.

Whenever a camera is used, record should be kept of the film type and speed, camera serial number, and name of photographer. Some police departments also require *f*-stop notation and recording of lighting used. These records may be important in courtroom presentation.

Proper evidence collection is dependent upon knowledge of what/how to collect, sufficient trained manpower, technical equipment available (e.g. photographic supplies, lighting or ladder if necessary, etc.), time to perform all necessary collection activities.[4]

Especially in cases of disaster and the large sites involved, there are a numerous materials of potential interest to forensic scientists. This can mean a very large number of exhibits to be collected, well beyond what is involved in a standard case. The key here is sufficient supplies (e.g. collection packaging), systematically searching the site and working according to established procedures.

[4] Problematic particularly in terrorist cases where there is a desire to restore normalcy as soon as possible, sometimes at the expense of proper evidence collection.

EVIDENCE COLLECTION KITS

Some items are relatively easy to retrieve, package and send for examination. It is a matter of packaging, labeling, and forwarding. An example of this would be pieces of a bomb after explosion.

Other items are large or bulky and not readily brought to laboratory. For this reason a series of *field kits* has been developed to aid investigators in deciding what items should be collected for laboratory examination. Some field kits are designed to select which persons handled explosives or weapons in order to single them out as possible suspects. A well-designed field kit might yield false positives (mistakenly suggesting, for example, that a given substance is explosive residue); it will not provide a false negative (erroneously showing that an explosive residue is from a harmless source). Results from field kit testing should always be confirmed by laboratory examination; these kit results should never be taken to court.

The Israel Police has designed a different type of kit specifically for disasters (Levinson and Amar, 1999). Rather than testing evidence, the kits developed by the Israel Police are best called *equipment kits*. They are designed to supply all of the equipment needed by each forensic and evidence collection function in a disaster. There is, for example, a kit containing all supplies needed for lifting fingerprints from the dead. Another kit has supplies for evidence packaging.

PURPOSE

The purpose of forensic laboratory examination is to select relevant evidence, interpret it, and establish what transpired, including the roles of both persons and materials. This totally objective process can assist the court to: (a) determine if the actions involved constitute a crime and (b) confirm the presumed innocence or criminal culpability of individuals.

Very often proof that a crime has been committed is based upon police forensic evidence. In one California air tragedy with all persons aboard killed, ballistics examination showed that the airplane pilot had been murdered (no one was left alive to testify or to prosecute). In the Happy Hour Supper Club fire in the Bronx, forensic evidence was used to supplement eyewitness testimony in proving arson as the cause of the incident.

Although the primary focus of forensic science is the courtroom, a by-product of the procedure is gathering knowledge that can be used in disaster prevention and mitigation. Perhaps some of the most commonly encountered examples are smoke detectors and a variety of equipment and methods to detect explosives and firearms.

LABORATORY OPERATIONS

The operation of a forensic laboratory in response to a disaster is little different from its operation in response to crime. Perhaps the primary difference is the large increase in the number of exhibits received. Basic operations still remain the same.

ASSISTING THE INVESTIGATION

There are cases in which forensic scientists have a particular direction in their findings, however the conclusion lacks the certainty required for courtroom presentation. In these cases forensic scientists can suggest to investigators possibilities of further investigation and possible/probable suspects.

DISASTER VICTIM IDENTIFICATION

In Western society it is accepted that the dead should be identified beyond all doubt, so that families can bring their loved ones to burial and reorganize their lives. In the context of disasters, this identification process is called Disaster Victim Identification (DVI). Note that it is not an immediate conclusion; it is often a lengthy procedure. The basic guiding rule is that making an error in identification is to be avoided at virtually all costs (including time delays).

It must be stressed that victim identification is critical particularly in Western society; this is not necessarily true in other cultures. Oriental religions place much less of an emphasis on identification, and in Eastern culture more circumstantial evidence of death is accepted. Islam stresses immediate burial after death, sometimes at the cost of complete identification of the deceased.

There have been numerous cases of dissatisfaction and criticism when Westerners have been killed in areas with a different cultural/religious approach to victim identification.

CONFRONTING DEATH

In Western society bereaved families are shielded as much as possible from the realities of death. Funeral parlors are designed to minimize the harsh reality as much as possible. Coffins have the elegance of furniture. At some viewings the deceased is beautified with make-up and clothing worthy of a social outing. Funeral services are also conducted in chapels that border on grand theater. Mourners and bereaved are comforted not only emotionally, but also physically protected from the pain and tragedy of death. Even our language uses polite circumlocutions – passed on, met his Maker – to avoid the directness of death. Rando (1984) has gone so far as to call the United States a "death-denying culture." When disaster occurs, the shock is all the greater. "Death denial" become a personal confrontation. All of a sudden the barriers of comfort and protection are lowered. Blood and suffering become all too close.

For the bereaved family, loss of a relative means separation from the deceased, transition into a new reality, and integration into a new life-style. It is not that the family forgets the deceased; they establish a new relationship with him (essentially with his memory). According to Rando most disaster response programs seal only the physical loss, and not its psychological aspects. Today society places strong emphasis on victim identification, to ascertain the identity of a deceased beyond all doubt. That identification is less and less by personal recognition, more and more by scientific methods. In other words the identification certifies physical separation and cancels all reasonable, unreasonable, and lingering hopes that the victim is alive – somehow, somewhere.

Victim identification is performed by forensic scientists (primarily but not exclusively forensic medicine). It does have, however, psychological importance. Identification is a process necessary to allow *closure* for affected families. This means psychologically coming to grips with the fact that someone has died. In cases in which no body has been recovered or identified, families have been known to reject even the strongest evidence of death. Nevertheless, when no body is found, it is incumbent upon responding agencies to record as much evidence as possible that proves death so that families can be brought to face reality.

Although *closure* is used as the term of accepting that death has occurred, in some cultures closure is never complete until the body has been found and brought to burial. Even when the family *is* convinced that a person has passed away, the absence of burial remains a bothersome factor.

Although in many jurisdictions there are provisions to declare a missing person dead after a given period of time, there simply are too many instances in which this has proven false, sometimes causing major problems (e.g. after remarriage of the spouse). To cite just one classic example, Henry Durant

(1862, reprint 1986, p. 48), founder of the International Committee of the Red Cross, records the case of an Austrian prince who was wounded in the Battle of Solferino, then treated by French surgeons. By the time he returned home, "his family . . . had given him up for dead, and had been mourning for him for several weeks."

When bodies are never retrieved, particularly when there is no conclusive proof of death, there still must be some form of closure on the part of the family.

As a result of an explosion of munitions at a factory in South Amboy, New Jersey (USA), 31 people died on 19 May 1950; most of those bodies were never recovered.

Pan Am 103 in Lockerbie is such an example. Although the plane crashed in an area that is quite possible to search (as opposed to the Thai International plane that crashed in the mountains of Nepal), the bodies of ten passengers and seven residents of the town (in the Sherwood Crescent area which was hardest hit by falling pieces of the aircraft) were never found. The working supposition is that they were incinerated in post-crash fires. The local police investigated the matter and provided proof of death.

Identification of a body also allows for the issuance of an official death certificate with the name of the victim. Further legal actions, such as the release of property for inheritance, insurance payments, and suits for damages, are based upon this document (Busuttil and Jones, 1992). There is usually considerable pressure that death certificates be issued reasonably soon after a disaster, so that families can organize their affairs and begin any legal procedures contemplated.

In the developed world very few disasters, except for civilian deaths in times of war, reach numbers beyond several hundred fatalities. In the United States only an extremely small number of incidents during the twentieth century included more than one thousand dead. For this reason there is a two-fold planning scheme.

SMALLER INCIDENTS

In the early part of the twentieth century private companies generally handled victim identification, particularly after isolated international incidents such as air crashes. The most prominent private company, dating back to the 1920s, was Kenyon Emergency Services, based in London (sold to international concerns in the 1990s). Particularly in the 1980s and 1990s many governments established capabilities to identify disaster victims.[5] The intent of all of these programs was to handle incidents with no more than a few hundred fatalities (in itself a tragic number). Very simply, larger numbers were just beyond the capability of private companies. Government declarations of ability were not as modest, since government agencies are well aware that the size of a disaster provides little justification to not take command, but effort was concentrated on the more frequent smaller incidents.

[5] One commonly cited disaster that served as a catalyst for government involvement was a propylene tank truck explosion in Spain on 11 July 1978, in which 150 people were killed.

The widest ranging government victim identification program to date has been set up under the US Aviation Disaster Family Assistance Act of 1996,[6] in the framework of which medical examiners can request assistance from the Department of Health and Human Services. This assistance takes the form of *D-Mort* (Disaster Mortuary) teams staffed with medical, technical and forensic members. The US Armed Forces Institute of Pathology has also been authorized to render victim identification assistance in civilian accidents. Again, the program has never been put into effect in a disaster with more than two or three hundred fatalities.

[6] The law originally applied only to US air carriers. An amendment to the law requires foreign carriers servicing the United States to file family assistance programs following air crash.

LARGER MASS FATALITY[7] INCIDENTS

[7] It is important to draw the distinction between mass *casualty* (injured) and mass *fatality* (dead) incidents. Each requires its own deployment. There is no reason to send a large number of ambulances to a disaster with no survivors. Many DVI personnel are not needed if there are no fatalities.

In earlier times there was less sensitivity to victim identification, and detailed reports of activities have not always been kept. In most cases, such as the Halifax Harbor (Nova Scotia, Canada) explosion (6 December 1917 leaving an official count of 1654 fatalities) and the 1900 Galveston, Texas (USA) hurricane and flood (6000 fatalities), any victim identification was made on the basis of personal recognition, sometimes supported by clothing and location of the body. It is true that these were large-size catastrophes that would overload any response system. The fact is, however, that there was essentially no victim identification system in place to be overloaded.

Until the latter years of the twentieth century planning for disaster victim identification – for disasters in general – was based upon battlefield models, usually taking into consideration civilian casualty experiences from the Second World War. These models were even extended to transportation-related incidents. Military personnel files contain *ante mortem* data. Military records can show quite accurately which soldiers are missing in action (MIA). In military incidents there is also strong control of the press, and families of soldiers are usually far away from the fighting. All of these characteristics are lacking in civilian models.

The largest numbers of Second World War civilian fatalities were on the Russian front (from where information and reports are missing) and in Nazi or distant cities such as Hamburg, Dresden (145,000 as a result of Allied bombing), Berlin, and Manila (40,000 dead when American troops entered the

Although the Halifax Harbor (Canada) incident can now be shown to have been an accident, at the time there was a strong public fear of sabotage or German military action. There were rumors of planned enemy operations against the port, even to the extent of testimony about supposed sighting of enemy ships. This was an *hallucination*, seeing something that is not there. A *delusion* is the misinterpretation of something that does exist.

This fear should be considered in the context of enforced night-time blackouts in Halifax. The explosions were so strong that they caused Richter readings in excess of 9 (Price, 1920).

Background: A munitions factory at Tom Black, New Jersey (USA) had exploded on 30 July 1916 (Dunn, 1917), with explosions continuing for 30 hours and shrapnel damage to the Statue of Liberty. It took considerable time before the heat of the fire was reduced and fire-fighters could start putting out flames. Unconfirmed rumors were rife regarding sabotage, although at the time of the Halifax accident proof was still lacking. Only in January 1940 were the rumors of sabotage put to rest, when German Capt Franz von Rintelen admitted that he had been the head of a 3000-person spying and sabotage network that had planned the Tom Black explosion. Von Rintelen confirmed that he played no role in the Halifax disaster (*Newark Evening News*, 3 January 1940).

city). One of the better documented models of Second World War civilian identification is Sheffield (UK), where 668 civilians were killed and another 92 reported missing and not found. Unidentified body parts were judged to account for 47 persons (among those missing). Initial body description and place found were incomplete, leading to a large number of bodies not identified and mass burials. Those identified by the Police Criminal Identification Department were buried in regular graves according to family wishes (Walton and Lamb, 1980).

Since peacetime disasters of this magnitude have not occurred in Western Europe, the United States and Canada, the basic application of Second World War-based thinking was to Third World American-sponsored assistance programs, where it was convenient to ignore victim identification as being a technical impossibility, much as had happened during wartime.

Only in the 1990s were there nascent efforts to change the conceptual bias against victim identification in assistance programs. Israel dispatched victim identification personnel to Turkey following the 17 August 1999 earthquake in the Western part of Anatolia; more than 16,000 persons are confirmed to have perished. The assignment of the Israelis, however, was to identify those Israeli citizens who were killed in the earthquake. Preliminary efforts were made to identify Turkish victims, however the effort was reportedly abandoned. After a plane crash in the Comoro Islands a similar Israeli DVI team was dispatched to identify Israeli dead. They identified numerous non-Israelis, but again the effort stopped, mainly due to bureaucratic complications.

On 23 November 1996 three hijackers commandeered an Ethiopian Airways flight to Nairobi. The hijackers demanded that the captain fly the aircraft to Australia. After unsuccessful negotiation the aircraft crashed into the Indian Ocean, about 500 meters from Great Comoro Island; 125 of the 175 passengers on board were killed.

A similar, if unintended, international effort to aid in victim identification occurred in the former Yugoslavia of the 1990s, as the uncovering of mass graves and the requirements of international legal tribunals made at least a minimal level of victim identification necessary.

There have been cases of competition and complaint when both commercial and government DVI teams have handled a disaster response. This happened in particular after one air crash in the Far East. One way to avoid bureaucratic competition is by the clear definition of roles to assure there is no overlap in responsibilities. Those definitions, however, are irrelevant in the post-incident reports of what happened, particularly when handling of the incident takes upon itself what could be cynically called a marketing commercial.

FRAMEWORK

The *definition* of identification is the positive comparison of *post mortem* (after death) and *ante mortem* (prior to death)[8] information of sufficient significance. The collection of personal *post mortem* information is the function of forensic medical personnel. The collection of *ante mortem* information is a combined effort of several different organizations, usually directed by the police.

[8] Time of death is called *peri mortem*.

Extensive professional literature is narrowly focused on the technical problems of recovering *post mortem* data from the body and making a technical forensic comparison with *ante mortem* records (e.g. Stuart, 1985). One of the most difficult aspects of victim identification, and also one that does not receive much attention in professional literature, is the *collection* of *ante mortem* information.

A perennial question is asked regarding what constitutes "information of sufficient significance." This is best answered by an expert, given the specific details of the case. One would not, for example, say that determination of a victim's sex is sufficient for identification. In one case in Israel, however, investigation confirmed a helicopter flight manifest showing the names of five people – four men and one woman – aboard at the time of crash. When the female body was recovered, gender was deemed sufficient for identification purposes.

Another version of this question is: what is the number of identification methods required to arrive at a definite conclusion of identity? In classic terms, if a victim is identified by his fingerprints, is it necessary to back this up with other types of identification, such as DNA or odontology to limit the possibility of error? If a victim is identified by dentistry, is a second method needed? Preferred methods of identification are odontology, fingerprints and DNA, but there is no generally agreed-upon answer if more than one of these methods is to be desired. Perhaps there is a pragmatic balance between the certainty of the identification and the ease/difficulty in obtaining further *post mortem* data.

There is also a question of principle in the determination of identification. In general, the police and the medical examiner present their findings in courts of law, and it is the court that either accepts or rejects the expert opinion. For example, eyewitnesses to a crime testify in court. It is not up to the police to certify the validity of their testimony – it is up to the court. It is ironic that in most jurisdictions the police and medical examiners certify the same eyewitness in cases of victim identification. The court is not involved. In recent years there has been a movement in Jewish law to have religious courts review and approve identifications of disaster victims. There is then an official court document which can serve as the basis for future legal action, such as remarriage or inheritance.

ANTE MORTEM

Problems of victim identification can be divided into cases with *open* and *closed* populations. In a closed population the names of all victims are known, and there is only a question of deciding which body belongs to which person. Military flights tend to have closed populations (as does a uniformed army, but larger). Flight manifests are tightly regulated, and there is no provision to allow "hitch-hikers" aboard; even military personnel using available space are part of the flight manifest. After a crash the problem is which body is to be associated with each of the names on the flight manifest.[9]

A commercial airline manifest can be considered a *semi-closed* population. The names of crew members might be exact, but passenger lists often have a margin of error, either due to last-minute changes or to factors such as spelling errors. Again, however, victim identification becomes the challenge of reconciling the passenger list and determining who is uninjured, who is hospitalized, and who is deceased.

[9] In a closed population it is possible to use *exclusion* as a method of identification. In the Israeli helicopter case female remains were immediately associated with the only woman aboard.

The collapse of a public building is an example of an accident with an *open* population. Virtually anyone might have been in the building. The same is true for a public bus or subway, with the exception of some long-distance buses on which passenger names are recorded.

The difference between open and closed populations is one of the primary reasons why military response models do not work in civilian situations.

Some *ante mortem* information can be found in general records. In Israel several bombing victims were identified by checking *post mortem* fingerprint forms against police files. US FBI fingerprint files are much more inclusive than files in Israel; they contain not only criminal records, but also other categories of individuals, such as those issued government security clearances. In one bus-bombing case the victim was not a criminal; his fingerprints were on file because he was a policeman. In most cases, however, it is necessary to establish missing persons lists;[10] these names become the focal point of classic-type investigations to obtain *ante mortem* information.

[10] For military situations the ICRC has a Central Tracing Agency to locate missing persons (military and civilian).

MISSING PERSONS

In an incident with a closed or semi-closed population, the list of people involved is usually known very quickly. With an open population the basis of compiling files is predominantly missing-persons reports, developed either through police questioning or reporting by the public. This is very different from ordinary police missing persons activity.

- In *routine* work missing persons reports are filed because there is no apparent reason for a person's absence.
- In *disaster* work it is the incident that often functions as a catalyst for people to file a report, fearing that someone *might* have been involved.

The net result is that in an open population disaster incident there will often be many more reports of people missing than there are people really missing.

In some cases missing persons reports should be taken with a certain skepticism. Following the AMIA building bombing in Buenos Aires one person reported himself missing as an ego ploy; another person reported missing was indeed missing – for more than 20 years! In the Bijlmer crash of El Al 1588 persons were reported missing on the ground; in the end only 43 were certified as dead. (Some observers cited the availability of government funds and the granting of residency rights to surviving relatives as incentives to file missing persons reports.) In more than one case, the "missing" person had not made alimony payments in more than ten years; police estimated that the missing persons reports were generated by the desire to try to force a more-than-routine investigation (Welten, 1993). Hodgkinson and Stewart (1988) also record this phenomenon of missing for many years; they interpret it as a search by the family for closure.

It is also rare that a family will cancel a report after the person is found safe; removal of the person from missing person rolls is done only as a result of police investigation or inquiry.

For these reasons missing persons lists from a disaster should be marked and not totally integrated into general missing persons files, even when the disaster case is closed and the missing person still unaccounted for.

POST MORTEM

Identification is a team effort. In some cases it takes workers of various disciplines to identify a body. In the dental area just one identification may require the coordinated work of a forensic odontologist, a photographer, a serologist, a forensic anthropologist, etc. (Brown and Powell, 1975b).

In a large operation work is segmented with different groups working on different identification methods (pathology, dental, etc.). The result is that bodies are often identified by more than one method (also encouraged by some as a cross-check against making mistakes). Statistics from Lockerbie illustrate this procedure. In total, 270 persons lost their lives. Of these, 253 bodies were recovered. Eighteen persons were identified by teeth only, and 13 were identified by fingerprints only. Seventy-eight, however, were identified by both dental and fingerprints. Numbers were not as definitive following the 16 August 1987 crash of NW 255 in Wayne County, Michigan, but the same was followed. Fifty-three people were identified by dental only, but 129 were identified by dental and at least one other method. (In total, 31 dentists worked on the identification operation.)

Very often identification errors are literally "buried," and no one is ever the wiser. That, however, is not always true. There are two types of error, sometimes hard to distinguish after a mistake has been made:

- Improperly identified
- Properly identified but improperly labeled

After the repatriation of one Lockerbie victim, the family wanted to have one last look at their relative before burial. When the casket was opened, the body was found to be of the wrong sex. As an inquiry later revealed, the identification had been correct, however there was an error in paperwork. The funeral was postponed until the correct body was obtained.

In the NW 255 crash there was an error that caused the exhumation of a body and its subsequent reburial under another name. The traumatic effect on the family is obvious.

HANDLING THE DEAD

There are many disaster response myths involved in the handling of the dead, particularly regarding infectious diseases. Very often the willingness to believe exaggerated claims of infection is part of the psychological phenomenon of avoidance, rather than any real medical danger. The "danger" is a reason not to become involved.

Hershiser and Quarantelli (1976) record that in their experience respect for the dead increases as the identity of the deceased is discovered. This identification is part of "humanizing" remains, particularly when one is talking about less than a full body.

RIGOR MORTIS

Stiffening of muscles (*rigor mortis*) is a phenomenon which can make examining a body decidedly difficult. The muscle stiffening is first perceived in face and neck about six hours after death, then spreads to the trunk and lower body. From 24 to 36 hours after death the phenomenon begins to lessen and finally disappear.

Given these facts, care should be taken not to force and break body parts if there is no reason precluding awaiting the end of *rigor mortis*. For example, the mouth and jaws will loosen after rigor mortis has ceased. It many cases it is possible to wait until after *rigor mortis*, rather that cutting the mouth (viewed by some people as a defiling of the dead).

INTERNATIONAL COOPERATION

There is a great deal of mobility in the modern world, hence in many types of disasters it is very common to find foreigners involved. When foreigners (or citizens resident abroad) are killed in a disaster, identification is generally possible only with the cooperation of authorities in the country of permanent residence.

- Current information (e.g. medical or dental records) needed to make an identification is usually found where a person lives.
- Older information (e.g. police or military records) might be found in a country of birth or previous residence).

In particular airline passenger lists can be misleading, listing the "citizenship" of the traveler, without noting that he was also a citizen of yet another country as well. This was true of three Israelis aboard TW 800 that crashed off Long Island. They were all listed under a second nationality, even though at least one lived in Israel where *ante mortem* files were compiled. Citizenship might yield some records, but it should not be confused with permanent residence (even illegal residence!).

In requesting *ante mortem* information it is best to work through official channels. If direct contact does not already exist, *bona fides* can be established either using embassies or Interpol as a conduit.[11] Placing a telephone request to an unknown police department abroad might work in novels, but it is not really an accepted procedure for passing personal information. It must be remembered that medical or other personal records cannot be released without proper authorization.

[11] Clark (1991) is critical of using Interpol due to delays in receiving information. It has been the author's experience that time of reply is very much dependent upon the specific NCB (National Central Bureau – the national Interpol office) involved.

Personal records should never be given to a private company such as an airline. They should be released to the police, which can then use the airline as a "mail messenger." The records remain police property. If a private company is hired by a police department to assist in identification, personal records should still be given to the police. The recipient police department can then take responsibility to share the information with the private company, conforming to any local laws and regulations. That police responsibility also includes eventual return of the records.

One of the major problems that occurs in international contact and cooperation is caused by different national systems and paperwork. Multi-language Interpol Disaster Victim Identification forms[12] are recommended for the international exchange of information to bring standardization to the process.

[12] Official forms are available in English, French, Spanish and Arabic. Non-official versions are also available in Dutch, Hebrew, Icelandic and German. They are all based upon the same numbered answers that are independent of the language of the form.

VISUAL IDENTIFICATION

Personal recognition is known to incur a significant possibility of error (Le Bon, 1897; Zugibe *et al.*, 1996). Gerber (1952) phrased this possibility very succinctly: "Nearly every one [sic] who has had experience with identifications made through sight recognition has one or more anecdotes to cite of incorrect identifications." Even so, personal recognition still enjoys popular credence beyond its scientific validity (Osterburg and Ward, 1992). It is simple and rapid, and the possibility of error is dismissed by convincing oneself that the witness is certain, so there is no possibility of error.

Levinsohn (1999) discusses problems in the condition of the face before personal recognition is at all practical. Usually one thinks in terms of an undamaged face when contemplating personal recognition. Levinsohn explains implications for recognition caused by degrees of damage and missing parts.

On 31 October 2000 Singapore Airlines Flight Number SQ 006 crashed after trying to take off from a closed and litter-cluttered runway at Taipei's Chang-Kai-Chek Airport; the pilot was detained for more than a month by Taiwanese authorities, who brought charges of criminal negligence. According to press reports victims were identified by personal recognition, although it was sometimes backed up with eventual DNA examination results (*New York Times* Internet Edition, 2 November 2000). (The element of personal recognition was not acknowledged by Singapore Airlines in correspondence with this author.)

A decision should be made if the condition of the body allows viewing, and if the face is sufficiently intact to yield a valid identification. ("Recognition" of a damaged body is clearly a source of error.) There certainly are instances in which families are allowed to "identify" a body, even though the legal identification is made based upon technical or forensic evidence. This "participation" in the identification process is done for their own peace of mind.

All this notwithstanding Kübler-Ross (1969) recommends that bodies be shown to families to hasten the process of *closure*.

The identification of living persons is made primarily on the basis of soft tissues (Brown and Powell, 1975a), which deteriorate after death. For this reason Jewish law[13] sets three days after death (one hour after removal from the water after drowning) as the upper limit for personal recognition.[14] After that time it is felt that there has been too much change in the face to allow accurate recognition.

After three to five days the skin begins to blister. A day or two later tissue swelling begins, bloating the face and abdomen, and causing the eyes and tongue to protrude. Other changes take place as a result of weather conditions (such as temperature and humidity) and the passage of more time.

Decomposition was accelerated following the Lauda crash in Thailand, when bodies were stored without refrigeration in heat ranging from 35°C to 40°C, and 70 to 80 per cent humidity (Geide, 1992). A video film taken at the Forensic Institute in Bangkok shows numerous flies and insects swarming over the decaying bodies.[15]

Cooling, such as in a mortuary refrigerator, slows but does not totally eliminate body deterioration. Following the Air New Zealand crash at Mt Erebus, Antarctica, the bodies did not decompose while the scene was being documented. As bodies were recovered, they had to thaw out from the Antarctic cold before they could be examined properly.

[13] The purpose of the authors is not to impose religious ideas. Rather, the authors want to sensitize responders to religious requirements so they can best serve the public.

[14] *Even Ha-Ezer* 17. The question of extreme temperatures (both heat and cold) is also raised.

[15] Footage taken by the German Bundeskriminalamt (BKA).

The process of personal recognition of a deceased person can be emotionally traumatic for the witness[16] as well as prone to error. The death of a relative is always hard. Even if the person has been suffering from a terminal illness, there is a hope that he will be spared just a little longer – another day, another week, another month. Psychological difficulties are often increased following unexpected death, particularly as is encountered in a disaster. These difficulties are again increased when more than one family or group member meets unexpected death. Two typical witness reactions with implications for identification are:

[16] In one disaster response in which one author participated, a young newly married husband became hysterical and violent when he saw the body of his deceased wife. Security officers together with a paramedic were required to restrain the man.

- rejection of the reality (hence rejection of the identification)[17]
- non-critical acceptance of the identification (sometimes without really examining the deceased)

[17] Phrased in professional terms, the author saw one family reject identification of a body based on a very unique tattoo because of color shade.

It should be ascertained that the person making the identification is not intoxicated. In one case the psychological reaction of a wife, upon hearing of her husband's death, was to drink. When she arrived at the mortuary to make the identification, she erred – in good part due to her intoxicated state.

Other sources of error can be not having seen the deceased for a long time, and only a passing knowledge of the person to be identified (including confusing him with another person).

When personal recognition is used for identification, it is recommended that prior to viewing the deceased the witness be asked to describe the victim in detail.[18] When the body is shown, the witness should then be asked to point out aspects that he described. This has the added benefit of ascertaining that the witness is really examining the body.

[18] Note that in certain religions there is fundamental objection to viewing deceased bodies without a specific purpose.

For numerous reasons, not only including possibility of error, bodies identified on site at the disaster should still be sent to the mortuary. Very often examination of wounds is needed for investigatory purposes.

There have been attempts to lessen the psychological strain of viewing a body by showing families either video clips or photographs of the deceased. In Lockerbie, for example, five Pan Am 103 victims were identified by "direct visual"; nine were identified by "visual from photographs." It has been the experience of this author that since photographs lack depth and vary in color shades, they pose additional problems and possibility of error in personal recognition.

BODY STORAGE

Following extrication and during the identification process bodies are best stored at 37°F (3°C). Only after identification procedures are completed should the body be embalmed, commonly done with formalin (a 4 per cent solution of formaldehyde); this allows body storage at higher temperatures.

FORENSIC METHODS

Forensic methods provide the possibility of identification without error. The problem is that forensic examinations take longer than personal recognition and require a trained and experienced staff.

FINGERPRINTS

Fingerprints are an excellent method of victim identification without error (Hazen and Phillips, 1982), however they are not necessarily a simple solution in all cases. Taking *post mortem* prints can be a challenging task that is not always possible (FBI, 1949/1968). Sometimes, hands and fingers are not even recovered.

Bodies and fingers can be in several conditions: freshly deceased, rigor mortis, early decomposition, advanced decomposition, desiccation, burnt, water-soaked.

The basic method of taking fingerprints from freshly deceased persons is to strengthen the fingers if necessary, then roll them against paper. Special commercially marketed kits have been devised for the purpose, including amongst other things: chemicals to inject for strengthening of the fingers, ink, curved metal "spoons" into which the fingers are placed, forms cut into strips to place in the "spoons."

In *rigor mortis* it is necessary to straighten the fingers (or to wait until the body itself becomes more flexible). In many jurisdictions court authority is needed to cut flesh from the fingers.

There are numerous technical methods used to take fingerprints from bodies in poor conditions. Most fingerprint technicians are trained to deal with latent prints at scenes of crime. They do not have extensive experience or training in taking *post mortem* prints.

There are "inkless" fingerprint forms on the market. They can be effective, but in many cases they do not provide the detail obtained through regular inking.

In other cases the challenge is to find appropriate *ante mortem* material. The simplest source of *ante mortem* fingerprints is when one has the "luck" to deal with a victim having a criminal record.[19] When formal prints are not available, personal property in the home is often a good place to look for fingerprints. Examples of possible property are a can of shaving cream or a checkbook.

[19] Some countries, such as the United States, take fingerprints for civilian purposes, such as part of the security clearance investigation. Other countries, such as Argentina, take fingerprints from all citizens as part of the national identification card. At one time France took one fingerprint on certain cards, such as alien registration, but no central files were kept.

FORENSIC PATHOLOGY

There are two reasons that autopsies are conducted: (1) to determine what happened, and (2) to identify the victim. In many cases this can mean a thorough pathological examination (autopsy).

It is general procedure to autopsy all bodies in an incident if facilities allow such. After the toxic leak in Bhopal that killed an estimated 2000 persons only random pathological examinations were made. After one autopsy determined that the victim had been murdered and added to the disaster victims, autopsies were conducted on all bodies (private correspondence).

Sometimes other considerations are taken into account at an autopsy. For example, when Saudia Airlines (Lockheed L-1011-200) crashed on landing at Riyadh Airport (Saudi Arabia) on 19 August 1980, the plane slid down the runway for 2 minutes and 40 seconds before stopping; then the aircraft burst into flames. One of the issues addressed in the autopsies was to determine if there was smoke in the victims' lungs (indicating that the passengers survived the initial impact – i.e. the seat belts worked – and died later from smoke inhalation).

In an accident in the United Kingdom autopsy again showed that the seat belts had functioned properly. The victims had survived the air crash, only to die in its aftermath. According to NTSB (2001) more aviation disaster deaths are due to impact than to fire. This is reportedly due to fire mitigation and suppression measures.

Prior to autopsy the victim should be photographed in the condition upon arrival in the morgue. Where relevant, trace evidence (e.g. fibers, possible explosive residue) is collected. Clothes are then described and removed without ripping, since they are potentially courtroom evidence. The clothes are also not washed.[20] The body is then cleaned and photographed once again.

[20] According to Jewish law the bloodstained clothes of a dead person are buried with the blood, hence they should be returned to the family unlaundered.

An external examination of the body is also conducted before autopsy, noting wounds and irregularities. Only after the external examination is completed is an intrusive internal examination begun.

Sometimes the pathological determination of prior medical treatment can be decisive in the establishment of the victim's identity. On the other hand, pathological information is often non-personal. For example, results can be that a woman had undergone a hysterectomy (thus excluding all persons who had not had such an operation). The frequency of medical treatment is cultural and age dependent. Helpern (1957, unpublished) cites an aviation accident in which almost half of 20 elderly females killed had had a hysterectomy. As a consequence of this frequency, pathology was used as a primary method to identify only one victim.

It should be recognized that conducting of an autopsy is not universally accepted. Jewish law, for example, places very strong limitations on autopsies (Mittleman *et al.*, 1992). Limitations, however, do not mean an absolute proscription. Very often the religious objections are misunderstood by bereaved families who are guided more by tragic emotionalism than by religious knowledge.

FORENSIC ODONTOLOGY

As other fields in forensic science, odontology is a relatively new profession that first began to develop in the final years of the nineteenth century (Luntz, 1977).

A key figure in the development of the profession is Oscar Amoedo (1863–1945) of Cuba and professor of dentistry in Paris for many years. On 4 May 1897 a flash fire destroyed the Bazar de la Charit, killing 126 persons. Later that year at the International Medical Congress in Moscow, Amoedo presented a paper concerning dental identifications of the bodies (Amoedo, 1897). The following year he published L'art dentaire en medicine legale, a 600-page treatise (Amoedo, 1898). Forensic odontology was on its way to being a recognized profession.

Forensic odontologists tend to be regular dentists who extend their expertise (an extension of the organizational typology suggested by Dynes [1974]). Many of today's forensic odontologists have gained status in this profession either through organizational membership or through operational experience (sometimes a single event that is resurrected time and again in conference lectures and journal articles).

Forensic odontological examinations can be divided into two basic categories: (1) comparative, and (2) reconstructive. The former category deals with determination if two dental records belong to the same person. The latter category deals with reconstructing the characteristics of the person, such as age and gender.

Harvey (1976) has reported that teeth are resistant to slowly applied heat, but sudden heat at high temperatures can cause disintegration. In terms of disaster, although the outbreak of fire may be sudden, the rise of temperature in the mouth is often more gradual, since saliva, jaws, lips and tongue can all serve a protective function. The result is that in a large number of cases the teeth remain and can be used for identification.[21] Thus, odontology has been a primary method of victim identification in disasters with estimates that teeth have accounted for up to 80 per cent of "difficult" identifications. This reliance on forensic dentistry, however, will probably decline in coming years due to the increasing importance of DNA.

[21] It is extremely rare that a fire will totally destroy a body. Generally bones remain, although they can become brittle. Teeth are one of the last parts of the body to disintegrate.

Forensic odontological identification is based upon *ante mortem* and *post mortem* comparison of features in the teeth and jaws. The possibility of dental identification is very much culture-dependant. In the case of Gulf Air 771 (crashed 23 September 1983 in the United Arab Emirates), very few of the Pakistani passengers aboard the flight had received dental treatment; in the 1969 air crash at Gatwick Airport, only five of 47 Asians had dental treatment (Clark, 1986). There can be no identification in those cases in which *ante mortem* records for comparison cannot be found or do not exist.

On 5 January 1969 an Ariana Afghan Airlines flight (727-113C) crashed on landing at London Gatwick Airport, Surrey, UK. The plane struck trees and a residence approximately 2.4 kilometers from the runway. Five of the eight crew and 43 of the 54 passengers were killed. Two people in the house were also killed.

Some forensic odontologists recommend removal (resection) of the jaws from the mouth (particularly in cases of burning, see Emson, 1987) to enable full examination. This can cause public objection, since intentional removal of body parts is forbidden in certain religions and cultures.

Dental records are kept on forms known as "odontograms".[22] There is, however, no international standard for these forms, and it has been estimated that more than one hundred versions and as many notation systems are currently in use worldwide. Since 1971 there has been a concerted effort to designate the FDI (Federation Dentaire Internationale) as an international standard. A popular option to FDI has been the Universal System. Yet another popular system is Zsigmondy/Palmer (preliminary development 1861; subsequently improved by Palmer).[23]

When x-rays (radiographs) are also available, they are preferred to notations, since they show better detail and are not subject to arbitrary interpretation systems. Clark (1991) also brings a case in which radiographs show that the treating dentist had made an error in filling out the odontogram.

It is now accepted practice in certain countries to mark dentures for identification. In recent years this has been extended to implanted false teeth.

As in general disaster victim identification work, a common problem in dental identification is finding *ante mortem* files. Records that are too old can pose problems, when extensive dental work was done in the intervening years. That is one of the reasons for which it is strongly recommended that a forensic odontologist and not a regular dentist handle these cases.

The finding of appropriate *ante mortem* records can be complicated and time-consuming. Following the Piper Alpha explosions off the Scottish coast at 2200 hours on 6 July 1988, Clark (1991) recorded that *after* the identity of the treating

[22] The first popular use of this term was by Interpol (Lyons, France) in the DVI forms it prepared in the late 1960s.

[23] The Palmer method designates individual teeth. The mouth is divided into four quadrants (upper left, upper right, lower left, and lower right). Each individual tooth in the quadrant is given a name.

dentist was known, it still took 15 hours to obtain dental records and identify the victim.

Many airlines now keep medical records of flight crew, included dental information. This is done as part of disaster planning with the realization that *ante mortem* dental records can be needed in the identification process (Keiser-Nielsen *et al.*, 1981).

Since the 1970s (Siegel *et al.*, 1977) there have been numerous computer programs to assist in comparing databases of *ante mortem* and *post mortem* dental records. The first well-known program was CAPMI,[24] developed by Lorton and Langely (1984, 1986). Another program was NOVA*STATUS (Solheim *et al.*, 1982). Tenhunen and Makel (1993) developed a Finnish system. As time progressed, these programs became more user-friendly. Glazer (1996) developed CAV-ID (Computer Aided Victim Identification). All of these programs have the general problem of requiring a significant amount of input time for the results achieved. The proliferation of dental odontogram forms causes obvious complications, when considering computerization for records comparison. In most cases it is more time-efficient to work manually. Following one air disaster a forensic odontologist inputted more than one hundred records and achieved one or two identifications; the procedure was certainly not time-efficient, and was done more for research than for operational considerations.

[24] One of the first civilian uses of CAPMI was following the 31 August 1986 air crash at Cerritos, California, USA (Vale, 1987).

There is extensive technical literature, particularly since the late 1960s, concerning forensic odontology. Much of that literature is "dated," given improvements in computerization and the advent of DNA that has taken a new importance. Much is also repetitive, describing the field and its very basic methods. Basic classical textbooks are Luntz and Luntz (1973), and Cameron and Sims (1973).

K. Brown (1984) cites dental identification in the "Albury Pajama Girl" murder case (September, 1934 in New South Wales, Australia) as an early example of odontology and the problem of error. The case was solved with the aid of odontology ten years later. Brown suggests that two dentists work together on cases as an identification cross-check.

Age can be determined by dental development. Methods are summarized in Clark (1992). Guidelines for dental identification examinations have been approved by the American Board of Forensic Odontology (ABFO, 1994).

DNA

DNA (deoxyribonucleic acid), the genetic marker found in nucleated cells of the body, was discovered by Jeffreys (Jeffreys, Wilson and Thein, 1985) as a scientific method of identification. Since then identification testimony based on DNA has been allowed in courts throughout the world. The first major acceptance of DNA as sole evidence was in R. vs Adams, (UK) Court of Appeal (Criminal Division), 26 April 1996. DNA analysis is alternatively called DNA profiling.

There are several methods for DNA profiling. Most common are:

- RFLP Restriction Fragment Length Polymorphism (earlier system using multi-locus and single locus probes)
- PCR Polymerase Chain Reaction
- mtDNA Mitochondrial DNA

The best source of comparative DNA is comparing *post mortem* samples with *ante mortem* samples from the suspected victim himself. In one case known to the author, a blood bank sample given by the victim was retrieved. In another case hair was taken from the victim's brush found in his residence. These samples have the advantage of high probability in comparison results.

In several cases known to the authors, families have refused to provide DNA *ante mortem* samples, in part due to fear of opening up family secrets regarding parenthood and brotherhood relations (or "non-relations"). Taking samples of the victim himself avoids this problem. These samples can be found on a comb, toothbrush, etc.

DNA collections can be found in both the civilian and military sectors. In some countries, such as Great Britain, DNA samples are now part of the criminal arrest record. The United States army started its DNA collection in June 1992 (Weedn, 1998).

Post mortem samples

There is no basic problem in retrieving a DNA sample from a more or less "complete" body recently deceased. In such a case body fluids[25] are available and can be sampled (taking into consideration, of course, approved medical safety procedures). The situation is much different with decayed or very partial bodies, in which case experts should be summoned to take samples. Smith *et al.* (1993) suggest that dental DNA is best taken by dentists. Others suggest similar specialized treatment for hair root collection.

[25] Common examples are saliva, blood, semen, hair roots, dental pulp and bone marrow. *Ante mortem* samples are usually saliva or blood.

Ante mortem samples

If the suspected victim's own DNA is not available as *ante mortem*, samples can be taken from parents, siblings, and offspring. In these cases a DNA expert should be consulted concerning which samples are most desirable under the specific circumstances. In any event, these samples will not provide results of the same level of certainty as the victim's own samples.

There are numerous kits available commercially to collect saliva (buccal) samples from family members. An example is the MasterAmp™ Buccal Swab DNA Extraction Kit (Epicenter Technologies, Madison, Wisconsin) that is marketed with extraction protocol.

Examination Procedures

The specific technical method of examination is at the discretion of an expert.

When DNA samples are shipped from one location to another, they should be labeled in such a way that the package need not be opened to obtain a full description of its contents. Thus, all administrative processing can be handled before the samples are opened under laboratory-controlled safety conditions. After all, DNA samples are biological material that carries with it the possibility of infectious disease.

In many cases in which there are dismembered bodies the question arises as to what extent DNA should be used to match remains. Although the theoretical possibility exists to sort all remains by DNA, the process can be extremely expensive and time consuming. There is, obviously, no one "right" answer. The issue very much depends upon cultural values and religious requirements.

In one air disaster body parts were divided into four basic categories: (1) pieces identifiable by means other than DNA (e.g. dental, fingerprints), (2) large pieces not readily identifiable by other means (e.g. torso, leg), (3) small fragments of flesh, and (4) tissues, fat, etc. Priority for DNA examination was given to the second category. After all examinations were made in this category, those category one cases that were still unidentified were submitted for DNA examination.

DNA examination can be time-consuming. In the case of the 1999 crash of EgyptAir, DNA examination took some 11 months (*Los Angeles Times* Internet Edition, 30 October 2000). In the Swissair crash DNA examination using Profiler Plus eventually identified pieces of 228 passengers (in total there were 229, including identical twins that could not be differentiated by DNA).[26] This, however, was done after some 50 workers ran DNA examinations on 2499 pieces of human remains (Kerr, 2001). The process took 105 days (more than three months) ([Toronto] *Globe and Mail*, 23 December 1998, p. A4).

[26] Identifications were as follows: visual – 1, x-ray – 4, fingerprints – 33, dental – 102, DNA – 89.

FORENSIC ANTHROPOLOGY

It is rare for a forensic anthropologist to identify a body based only on his own work. On the other hand, the anthropologist can be an important member of the identification team. Contributions of the anthropologist can be:

- Determining whether remains are human or animal
- Sorting of commingled skeletal remains

- Facial reconstruction from skeletal skull
- Photographic superimposition
- Determination of sex
- Estimations of age, height, race
- Possible disease (from certain bone damage)
- Determination of race

Just as estimated age can be determined from fetal or dental development, it can also be determined by skeletal development.

Race determination is limited to Caucasian (narrower faces with high noses), Mongoloid/Western Hemisphere Indian (forward projecting cheekbones), and Negroid (wider nasal openings and sub-nasal grooves) (Mann and Ubelaker, 1990).

On occasion facial reconstruction has been tried to assist in victim identification. When this is done, it should be remembered that the method has serious limitations. Reconstruction deals with bone structure, yet most recognition is done based upon soft tissues and skin. This should be taken into consideration before a reconstruction is undertaken.

MISCELLANEOUS

PODIATRY

At times it is desirable, if not necessary, to consult with very specialized resources. Forensic podiatry, for example, was first cited in 1935 (Muir, 1935), given the uniqueness of the foot and its identifiability. Doney (1984, unpublished conference presentation) and Vernon (1994) both suggest that forensic podiatrists can identify foot treatment and unique lesions (e.g. haemanigioma, cyst, pigmented naevus).

TATTOOS AND BODY PIERCING

The use of tattoos is very much cultural. In some societies tattoos are extremely rare. In Jewish law, for example, the application of a tattoo to one's body is forbidden. In other societies, primarily maritime-oriented, tattoos are much more common. In the Piper Alpha disaster (explosion on an off-shore oil rig), for example, 24 of the 136 bodies recovered were identified by tattoos (i.e. 17.6 per cent) (Clark, 1991).

A permanent tattoo on a person is the result of injecting ink three layers beyond surface skin. This can be done with either hobbyist or professional equipment. There are two basic types of tattoos: (1) standard (e.g. given to a large number of seamen serving on a particular ship), and (2) individual (designed for a particular person). Obviously, the latter type provides a much better basis for identification.

ANCILLARY EVIDENCE

PERSONAL PROPERTY

It is best not to base an identification of a body upon property, however it is recognized that under certain circumstances such may be necessary. In any event, property (including both clothing and belongings) can be an important investigatory tool in suggesting an identification and the resultant search for *ante mortem* information.

It should be noted, that property (including clothing) is not always to be found on a body. One case is Pan Am Flight 103 (Lockerbie). The airplane exploded roughly 11 kilometers in the air. During the approximately 2.5 minute fall (sometimes exceeding 120 miles per hour) most clothing, except for tightly tied items such as shoes, was ripped from the victims. In other cases, such as the *Mineral Dampier* sinking, the victims were not wearing much clothing or carrying extensive property at the time of the disaster (Kahana *et al.*, 1999). This lack of clothing and belongings can complicate the ensuing scientific identification. In the *Noronic* incident, virtually all property was destroyed by fire (T. Brown, 1952).

The property found on a body can be divided into two major groups: (1) items loaned to other people, and (2) items not loaned to other people.

The position should be noted of property found: (a) on, (b) next to, or (c) near a body, since the item in question does not necessarily belong to the cadaver. A case in point is a national identification card found atop the body of a deceased victim following a bus tragedy in Israel on 6 July 1989. Although the identification looked straightforward, it seemed odd that the identity card was found on the victim, rather than in his pocket. As the matter turned out, the card did not belong to the victim at all. It had fallen out of the pocket of a volunteer responder.

In another instance, at 0811 on 5 October 1999 at Ladbroke Grove, a train from Cheltenham *en route* to London–Paddington Station collided with a local Thames train bound for Bedwyn, Wiltshire. Fire engulfed one of the cars leaving 31 dead. Evelyn Paler climbed out of the wreckage of the Paddington rail crash and draped her coat around the body of a dying woman. Disaster responders later searched the coat, found work papers and a train pass, then notified the family of Ms Paler to expect the worst. Fortunately, the mistake lasted only for several minutes until Ms Paler called home to say that she was safe.

It is rare that a victim should be identified solely on the basis of property. This, however, is the professional decision of those persons charged with the responsibility of identification. Items that carry heavy weight are expensive jewelry, inscribed items, travelers checks, etc.

Most people traveling, whether by air or bus, carry cash, sometimes in significant amounts. It has been known the surviving families have filed claims of monies purportedly carried that well exceeded what was found. For this reason all property should be logged by at least two persons. The listing of money often has the problem of partially burnt notes after a fire. If these cannot be properly recorded, photocopying is a frequent alternative.

It is insufficient to search for cash in pockets and concealed money belts. Many people are afraid of pickpocketing and theft, as a result of which they hide money in unusual places. In the Air Florida crash one victim hid $1600 in her sock!

MISCELLANEOUS

Airplane seating charts cannot be relied upon for accuracy, since passengers often change seats once aboard. In England the names of passengers are often recorded on inter-city buses; these lists also are not always accurate.

BODY REPATRIATION

The return of a deceased victim to his country of origin is usually called *repatriation*. An experienced international undertaker generally handles the preparation for body shipment. This includes the drawing up of proper paperwork, embalming as required by the country of receipt, and the use of a coffin meeting international standards. The transportation carrier on which the victim had been a passenger generally hires the undertaker for this purpose.

Families should be offered the choice of repatriation, local burial in the region near the disaster site, or cremation. There have been instances in which family relationships are not the most amicable, and local burial is chosen. One example is a surviving wife who did not want to receive her husband's remains due to bitter divorce proceedings. Another example is a second wife who wanted a quiet burial far away from the first family. There are also some countries that do not allow receipt of deceased victims.[27] Thus, there is no option of repatriation.

[27] When it was realized that only deceased persons would be recovered from the *Mineral Dampier* sea disaster, the retrieval operations base was moved to South Korea, since those bodies recovered could not be brought into Japan.

In various cultures cremation is preferred. In some countries, such as Israel and numerous Moslem countries, cremation is against religious law, and there are simply no facilities available to cremate a body despite family wishes. In a *cremation* the body of the deceased is taken to the crematory. In some jurisdictions there is a period required by law before cremation can be done. The actual cremation usually takes two to five hours, as the body is burned at a temperature between 760ºC to 1150ºC. It is not only burning; hard items, such as teeth, are often pulverized. The final remains usually weigh between 1.75 and 3.7 kilograms; they are placed in a container, which can be buried or otherwise retained by the family.

THE MORTUARY

Disasters often overwhelm existing mortuary facilities, thus requiring the establishment of temporary mortuaries created for the situation. These temporary facilities most often house the identification and forensic medical operations. (Since police investigations are so closely tied to the victims, it is common that the investigations center be located in/near the temporary mortuary, at least in initial response stages.)

Lockerbie, under usual circumstances, is a quiet Scottish town. Mortuary facilities are minimal. After the crash of PA 103, Taft and Reynolds (1994) report that no fewer than 190 policemen staffed the mortuary operation in the town. The incident clearly overwhelmed the relatively minimal capabilities of existing funeral homes. Bodies were brought to the town hall, the ice skating rink, and warehouses in the town.

In the response to the Air Inter air crash (20 January 1992 at Mount Sainte Odile, France) bodies were stored in a temporary mortuary (tents) in the town of Barr, then for examination they were transferred in small groups to the forensic mortuary facility in Strasbourg, a half hour drive away.

Many mortuaries and funeral parlors, particularly away from larger cities, are built for a low volume of work. They tend to be under-utilized and are also built to insulate the public from the realities of death as much as possible (Blashan and Quarantelli, n.d.). In these smaller mortuaries most examination procedures are conducted around a single table. In a disaster these mortuaries are often inappropriate. Disaster work is most efficient to work according to the station method, assigning to each of numerous tables a specific function (e.g. fingerprints, dental), then moving the bodies from station to station (rather than bringing the various experts to the body).[28]

The most practical temporary mortuaries are in facilities relatively simple to protect from public entry and view. This has often resulted in choosing buildings on military bases. Another option is a sports stadium, built to exclude non-paying fans and to facilitate police control.[29] Following the "Holiday on Ice" disaster in Indianapolis the ice-skating ring was used as a temporary mortuary (Blashan and Quarantelli, n.d.). The same type of facility proved problematic in Lockerbie, where bodies were placed on the ice at the skating

[28] At the mortuary facility in Dover, Delaware (USA) ten stations were set up to handle the 256 bodies (248 passengers and 8 crew) retrieved in Gander, Newfoundland following the 12 December 1985 crash of a DC-8 US Army contract flight (Arrow Air) (Clark *et al.*, 1989). These ten stations dealt with: control number and photography, inventory of personal effects, fingerprints, dental x-ray, general x-ray, dental examination, autopsy, embalming, body preparation, bodies into caskets. Another description of the operation at Dover can be found in Thompson *et al.* (1987). The exact number of stations is flexible. The above should be considered as only one suggested model.

[29] Ebbets Field, the home stadium of the Brooklyn Dodges baseball team, was used as a temporary medical treatment center for the same reasons after the 1 November 1918 subway crash in the Malbone Street Tunnel.

rink, but then they froze. The eventual solution was to place the bodies on skids, thus increasing the temperature sufficiently to avoid freezing.

At 2306 on 30 October 1963 during the final moments of a "Holiday on Ice" performance, a large explosion ripped through the Indianapolis, Indiana (USA) Coliseum on the Indiana State Fairgrounds. Fifty-four people were dead at the scene. Another 27 died as a result of their wounds. Almost 400 others recovered from their wounds. In all 4327 people were recorded as having been in attendance.

BEREAVED FAMILIES

A well-designed mortuary in regular use is built to shield the family of the deceased from the harshness of death. The atmosphere is quiet and dignified. Many times there is a separate area for family during the time of the funeral, so that other attendees do not see their grief. This is all missing in a temporary mortuary, where viewing of the body is performed as part of the identification process, without the practiced verbal skills of an experienced funeral director. In other words, the established mortuary is built to help the family cope with grief; the temporary mortuary is "work efficient" and not oriented towards coping with grief.

Grief is a normal reaction to sorrowful events such as death. Lindemann (1944), while discussing the Cocoanut Grove fire, states that even *acute grief*[30] is normal. Peschel and Peschel (1983) describe the stages of grief as numbness, disbelief and rage. Rando (1984) sets the cycle in more active terms: avoidance, confrontation, and re-establishment.[31] Valent (1995) describes depression as misdirected grief; it can obscure the initial manifestations of PTSD and their diagnosis/treatment.

External actions, such as making funeral arrangements, do not necessarily mean that in Rando's terminology the bereaved family member has progressed from avoidance to confrontation. That family member can still be in the first stages (Peschels' numbness) and be going through physical, but not psychological, actions. This is particularly true after disaster, which often presents cases of sudden and totally unexpected death.[32]

This author participated in one extreme case of this nature. Notification of death was made to the mother of the deceased boy, and the woman signed on paperwork. Despite her signature, the woman remained in the *avoidance* stage. In this case avoidance went to the extreme of *denial*. The following day

30 Today that acute grief can be called PTSD.

31 With the possibility that stages overlap.

32 According to Glick *et al.* (1974), sudden loss does not increase grief.

she filed legal papers and claimed that she knew nothing of her son's fate.

Denial is a known phenomenon in which people try to convince themselves that something negative did not happen. The above example is extreme. In most cases denial is manifested in the false hopes a family nurtures that a particular person is still alive, despite probable information to the contrary. After death notification is given, survivors often combat denial and try to come to grips with the death by discussing how the person died.

Grief does not begin when a family is confronted with the death of a loved one; it starts much earlier. Engel (1964) explains that *anticipatory grief* begins when the family realizes the severity of the situation. In a non-disaster situation, anticipatory grief can begin when a family is notified that someone is suffering from a terminal disease. Even though the sick person might live for months, the grief process (sometimes quite severe) already begins.[33] Sometimes that process is compressed, for example when a doctor starts hedging words as he reports on the results of surgery. In disaster terms, anticipatory grief can begin when family members realize that a loved one was *probably* involved in a fatal incident.

[33] Glick *et al.* (1974) report that the optimum period of anticipatory grief is no longer than six months. There have been reports of divorce after an extended period of anticipatory grief, since the spouse already "writes off" the ill partner.

Given these study results it would seem that psychological intervention is a possibility even in the earliest stages of anticipatory grief. In any event, no person given a message of death or serious injury should be left alone until it is ascertained that he is capable of coping with the news.

One method of coping is through ritual, whether it be religious or secular (e.g. military). Ritual provides the bereaved with a framework in which he can develop a new relationship with the deceased. In this sense, funerals are rituals designed to channel expressions of grief. The same is true for memorial services held both in the month or so after death and in years following. There are numerous other methods of coping.

The worst subway disaster in New York history occurred on 1 November 1918, when an inexperienced driver on the Brooklyn Rapid Transit (predecessor to BMT) Brighton Line passed a curve at excessive speed and derailed his train, crashing into a wall in the Malbone Street Tunnel, killing 94. As a coping method, officialdom tried to *disassociate* the area from the crash. Malbone Street was renamed Empire Boulevard.[34]

[34] A similar name change was made following the Great Northern Railroad crash near Wellington, Washington. The town was renamed "Tye."

BURIAL

In Western culture there is generally strong objection to mass burial (Hershiser and Quarantelli, 1976). Even when bodies are not identified, there is a general preference to bury the dead in individual, albeit anonymous, graves. In certain non-Western societies and under unusual circumstances there is less opposition to *mass graves*. Following the Bhopal disaster, for example, people were not only buried in mass graves; already closed graves in the local Moslem cemetery were reopened to allow emergency inclusion of additional dead (Silverstein, 1992). Many victims of the January 2001 earthquake in El Salvador were also buried in mass graves, reportedly up to 400 meters in length (*Washington Post*, 16 January 2001).

The disapproval towards mass graves also applies to wartime. One of the better documented models of Second World War civilian identification is Sheffield (UK), where 668 civilians were killed and another 92 reported missing and not found. Unidentified body parts were judged to account for 47 persons (amongst those missing). Initial body description and place found were incomplete, leading to a large number of bodies not identified and mass burials. Those identified by the Police Criminal Identification Department were buried in regular graves according to family wishes (Walton and Lamb, 1980).

An exception is unidentified smaller parts of bodies such as skin fragments and blood[35] that are often buried unobtrusively in unmarked mass graves.

Contrary to this policy of privacy, in the case of the Egyptair crash, unidentified body remains (not whole bodies) were buried in a *public* ceremony in a Rhode Island cemetery on the first anniversary of the crash (*Los Angeles Times* Internet Edition, 30 October 2000). This ceremony was thought to bring closure to the bereaved families and to mark the end of the case for Egyptair (except for the litigation that obviously continued).

[35] Disposal of blood is an unclear issue. Some religions (such as Judaism) require that it be buried. Other religions and cultures do not have special requirements, in which case blood is disposed of in accordance with procedures for biological waste.

BODIES NOT RECOVERED

There are numerous cases in which all bodies thought to have been involved in an incident are not recovered. This is much less frequent today than in the past, since DNA can be used to associate a body (once unidentifiable) with a deceased person. The problem, however, still does exist, as in the case of airplanes that crash over water (e.g. the TWA crash of 17 July 1996 in which not all of the 230 bodies were recovered). In an aviation accident, for example, proper handling of the case requires basic investigation proving that the person involved had been aboard the plane. This can include taking testimony from family members who might have taken the person to the airport, interviewing airline check-in staff, examining tickets and boarding cards, etc. This proof is required not only for legal reasons. It also leaves the family with a basis for closure.

The importance of this search by families for closure cannot be underestimated. Until there is definite proof of death (most preferably by identifying a body), families can nurture false (if not absolutely absurd) hopes that the person is still alive. Care should be taken, however, that families do not prematurely identify bodies (i.e. the wrong bodies) to hasten closure, or that they do not experience rejection and deny a scientific identification (particularly due to changes in the body from the time of death).

MISTAKES

Mistakes in identity should never be made, but in fact there have been cases in which mistakes were made. The cause is very rarely scientific error. More often mistakes can be traced to personal recognition or to administrative error, such as the mix-up of paperwork in either the identification or preparation-for-burial stages.

The most serious mistake to make is in the identification itself. There have been cases in which a body was thought to be that of a person still alive and well. The ramifications are obvious, as a family mourns someone who will yet come home. Less serious, yet emotionally very traumatic, is the mixing up of bodies before burial, such as the Lockerbie incident mentioned earlier. The family opened the casket and found that the body was of the wrong sex. There are no long-term ramifications here. Both people identified were, in fact, dead. Nonetheless, uncovering the unintentional switching of the bodies was an extremely difficult experience that ran the risk of undermining the credibility of other identifications.

In the Gulf Air crash of 23 August 2000 (Airbus A320-212), all 143 persons aboard were killed. Three Egyptians were misidentified and as a result mistakenly buried in Bahrain. This was "corrected" by exhumation and reburial of two of the three in Egypt. Although such exhumation was done quickly in this case, there are jurisdictions in which removal of a body from a grave can be a complicated and drawn-out legal procedure. There are also religions that place restrictions on such a practice and prefer relabeling of a grave rather than body exhumation.

Another problematic area is the association of body parts for the purpose of burial. In theory, DNA profiling can give definitive answers if two human fragments belong to the same body. The process, however, is impractical on a routine basis (particularly in a large disaster) due to time, cost, and manpower/equipment considerations. In professional terms, burying the wrong piece with the major part of a body has no real consequences, if the piece is not needed for another identification. Surviving families, though, do not necessarily see the matter in such objective and non-emotional terms.

Sometimes the situation, itself, is unusual and encourages error and confusion. After the snow avalanche of 1 March 1910 in Wellington, Washington (USA), 97 people were killed in two different trains belonging to the Great

Northern Railroad. Two different bodies were identified as John Brackman. Mistake? No. They were two brothers whose father, also John Brackman, gave them the same name. Claiming the body became more problematic. Each was wealthy and left a sizeable inheritance. There was no shortage of "next-of-kin" (Moody, 1998).

Hodgkinson and Stewart (1988) report that viewing of the deceased helps complete closure and allows for greater peace of mind. This is, of course, contingent upon the condition of the body. It also contradicts traditional advice to "remember the person as he was." In family terms this desire to see the deceased one last time is often phrased as the desire to "identify" the remains (often despite their condition).

FRAUD

One unpleasant reality is that there have been instances of fraud and deceit in victim identification. After the 1994 bombing in the AMIA building in Buenos Aires, one "widow" claimed that an unidentified body was that of her husband, then she filed for compensation and received $55,000. In 2001 the "deceased" was found alive, well and employed in Paraguay, where he was arrested (Reuters, 21 April 2001). This is an example of a problem involved in accepting verbal testimony without scientific verification.

CONCLUSION

The use of forensic science in incident investigation is an example of using modern technology to gain a maximum of information regarding what happened, and who was linked to the event. This information can be of benefit both in police investigations and in subsequent courtroom prosecution. All those expected to be at a disaster site, both professionals and volunteers, should be taught about evidence preservation during their contingency training.

CHAPTER 21

VOLUNTEERISM

INTRODUCTION

Volunteerism has always been a cherished value in virtually all societies. In times of disaster volunteers can be divided into two basic categories: (a) *spontaneous* responders who just happen to be at a disaster site, and (b) those who contribute their service *in advance* to assist official or recognized agencies in time of need. Volunteers in the latter category are perceived either as individuals helping an agency, or as an organization (e.g. the Red Cross, Israeli ZAKA,[1] or the Salvation Army) helping others.

People volunteer every day, to the extent that many public services are virtually dependent on people who step forward to provide predesignated services. One small example is hospitals, where people frequently volunteer to give social support to patients and their families. These efforts are well controlled, with volunteers working according to a schedule with assignments. This can even extend to "uniforms," such as those that some "candy stripers" wear. Volunteers also "sign in and out" for their assignments. Many ambulance services are heavily dependent upon volunteers for auxiliary staffing.

[1] In Israel the ZAKA volunteer network was established following a 6 July 1989 terrorist incident to clean blood and pieces of human remains from disaster sites. The network went through several types of organization and leadership until a more or less permanent model was established in the second half of the 1990s.

VOLUNTEER OPERATIONS

Running a volunteer program is not without its problems. Salary can function as a control over paid employees. They must conform with certain job norms to continue their employment. That control does not exist with volunteers. It is true that those working for a volunteer organization must adhere to instructions, but the leeway is much greater and there is more tolerance.

The Sahuarita, Arizona (USA) train derailment caused an interesting situation. The volunteers simply did not recognize a new reality in their status as a result of a bureaucratic change. The local jurisdiction was organized as a municipality in 1994, and the police was established three years later. Until that period volunteers to the regional Sheriff's office held responsibility for law enforcement. In the derailment response the volunteers, who "have been in this area 'forever,' . . . tried to take over. They were promptly put in their place as 'support' . . . [They] promptly lost interest in assisting, but stayed at one location, the Command Post" (personal correspondence).

It is not unusual that volunteers forget their status and position. In one case in Israel the leader of a volunteer group enlisted to *aid* the police complained in a cellular telephone conversation to a friend during the response, "The police are getting in our way!" (personal observation).

Another variation of this phenomenon is *emergent leadership* amongst volunteers, just as it exists with regular workers. What often happens is that a volunteer plays an unexpectedly prominent role in a disaster response and tries to continue in his high profile position, not realizing that the response bureaucracy "filled the gap" – in other words, designated a regular worker to perform the task done by the volunteer.

CONTROL

At a disaster site it is not possible to exercise complete control over volunteers. The natural reaction amongst many people is to want to help. The number of volunteers, in as much as people will step forward to help based on personal desire, is often divorced from an analytical assessment of objective need. If anything, the incident, itself, is the control on excess. When people see that victims are being cared for, most will not feel the need to step forward. In this sense, volunteer organizations are more disciplined, however there have been instances in which they, too, swing into action based upon subjective feelings rather than objective evaluation. Although difficult to say, the authors know of one volunteer organization who inflated their response for statistical purposes.

The net result is often the complaint of "too many" volunteers cluttering a disaster site. Given proper planning the right number of volunteers can be a positive part of an emergency response (although some difficulties will still remain). Without that planning they can constitute a problematic group and an almost unrivaled hindrance.

The Cypress Freeway collapse after the Loma Prieta earthquake is just one example of volunteers moving from being critical to life-saving operations to "a distraction" and "a problem" (Stark, 1990). The phenomenon is no different in transportation-related disasters.

SELECTION

A demographic analysis of the jurisdiction can be a key to the potential of recruiting volunteers. For example, people with rigid work schedules such as production-line factory workers have less chance to volunteer during working hours. They are, however, more available after work hours than an executive-type person whose perceived essential work extends to evening appointments. Older people, even when they do volunteer, can only be given certain types of jobs due to possible physical limitations. New arrivals in a neighborhood can have a lesser awareness of local problems, or time demands on their adjustment to the area that preclude volunteer activities.

In a volunteer program not everyone who steps forward is suitable. The same, of course, is true for the general job market. Although slightly more leniency can be allowed in recruiting volunteers, it should be remembered that training costs money, equipping costs money, and operating costs money. Volunteerism is absolutely not "cheap labor" as is often contended. There is no justification to invest in personnel who are not suited for the job to be done. Disaster-related work can also carry with it significant psychological burdens. This should be taken into consideration when recruiting volunteers.

Depending upon the task in disaster response, physical fitness can be a definite factor in recruiting volunteers. Many disaster sites are dangerous because of broken and bent metal, heat after a fire, or oil/fuel spillage. Psychological strength is another personal factor that should be taken into consideration.

Age should not be a factor in recruiting volunteers, although it is a consideration in their assignment. As a general guideline, volunteerism is heaviest between the ages of 25 and 45, when it tails off. Personal experience has shown little long-term volunteerism before age 25 or after age 60. Where such volunteerism does exist, its practical value is often questionable. Care should be exercised in using young volunteers (e.g. high school students) in disaster response situations involving difficult sights (victims, bereaved families, destruction, etc.).

In selecting volunteers, their outside commitments should be taken into consideration. A mother with young children at home might find it quite difficult to volunteer when an emergency situation takes place during the day when her husband is at work. Someone whose workplace is at a significant distance might show up at a disaster after the response is over.

PROBLEMS

Today it is impossible to run any planned program without due regard to questions of risk and insurance. Coverage should be arranged for volunteers during the hours when they are activated, from the time of call-up though assignment completion. It should be remembered, however, that certain disaster-incurred physical and psychological problems only become evident at a later stage, hence there is a delay in their being reported. This has to be taken into consideration in writing any policy.

There are many problems inherent in running any pre-planned volunteer operation. On the very personal level the volunteer might constitute a decided organizational benefit, but paradoxically he is often regarded with distrust, if not contempt, by the paid worker who receives a salary to perform tasks the volunteer performs without payment. This problem does not exist, of course, in *ad hoc* life-saving situations, when performing psychologically difficult tasks, or when the actual service rendered by the volunteer does not exist in the responding organization.

There can also be a high turnover of volunteers, especially in disaster response. Some people find that as work progresses the tasks involved in volunteering, especially in a disaster, are "not for them." Many volunteers require some action to achieve satisfaction, and since disasters (hopefully) do not occur very often, many volunteers find other pursuits.

Israel was confronted by a wave of terrorism in the mid-1990s. When terrorist incidents ceased, the members of one disaster volunteer organization needed other directions in which to exert their energies. They turned their focus to assistance in crime and traffic deaths. Despite the close connection of the volunteers to the Israel Police – even their induction into the Israel Police Civilian Guard – they never mastered police thinking and priorities. This became a constant source of friction.

Volunteers must be understood before they can be handled properly. A good part of this understanding is connected to their motivation.

MOTIVATION

In times of need there is a basic human desire to want to help. Volunteers are usually defined as those who step forward to help "without compensation." Nothing could be further from the truth. Although the volunteer waives monetary recompense, he is definitely looking for compensation – albeit in such areas as public recognition or "fame."

Thus, it is a mistake to think that wanting to "do good" is the only goal of volunteers. The act of volunteering is the result of a number of motivating factors combined. It is this "package" of factors, some more important and some less important, that motivates volunteers. These factors also change, both in composition and relative importance, as a disaster incident progresses. The motivating factors include:

- Desire to help
- Religious or moral duty
- Curiosity
- Mingling with "important" people

- Excitement
- Need to be "important"
- Peer pressure
- Need for recognition
- Publicity
- Access to information

Although the unemployed might well have the spare time to be involved in volunteer activities, it should be realized that the satisfaction from these activities cannot replace the satisfaction enjoyed from gainful employment (Reinholtd, 1999–2000). In the case of short-term unemployment, volunteerism has the danger of being a transient and temporary activity. In cases of long-term unemployment, filling the time void with volunteer activity can pose the psychological crisis of acknowledging defeat in the job market. After all, volunteerism under these circumstances can be perceived (whether rightly or wrongly) as accepting work *without pay*, because there is no work *with* pay.

In one case a very negative aspect of "volunteering" was reported. As one person was dressed as clergy (which he absolutely was not) and consoling families of fatalities, his cohorts were busy robbing the houses of the bereaved. The lesson is obvious. The *bona fides* of volunteers with access to information or sensitive subjects should be verified.

Volunteerism often starts with convergence. This is the natural tendency of many people to converge or gather around a disaster site, if for no other reason than to find out what happened. At this initial stage there will be those who rush to help (become involved), and those who stand to watch or observe. In this stage of an incident volunteer behavior tends to be perceived as primarily altruistic. Those who move forward to help are generally motivated by a genuine desire to "do good." As has already been pointed out, it is a fallacy, however, to think that this is the only motivation. Even at the initial stage other motivational factors are valid.

As professional response officials begin to enter the disaster site, the need for untrained volunteers to "do good" decreases, yet it is common that the volunteers try to remain at the site. A small number of the volunteers will continue to render service, changing their roles from life-saving tasks to giving assistance to life-savers. Others will slowly join the convergence of spectators. (The perceived status of a spectator/volunteer within the police barrier is considered higher than the status of general public spectators. Tape is a psychological barrier closing off an area; metal dividers are a more physical demarcation. A person inside a metal barrier is of higher perceived status than a person inside a tape barrier. Following one bus bombing in Israel an absurd situation developed in which spectators/volunteers with no official status dictated who would be allowed inside the incident barrier to render assistance!) With this change of role there is also a change in motivation. At this point less idealistic motivators (such as importance or excitement) take on more importance.

It should be very clear that the motivational factors listed above are not necessarily limited to volunteers. Many professional responders choose their jobs for some of the same reasons; employment simply "legitimizes" these motivators. Thus, there is the odd paradox that a motivator such as "excitement" or "importance" is considered a negative reason to volunteer, but a perfectly acceptable reason to accept a particular job assignment.

It is not surprising that during an incident even regular response personnel exhibit some of the same traits as volunteers. After the series of bus bombings that plagued Israel in the mid-1990s many of the technicians continued to be present at the morgue long after their official tasks were completed. Perhaps the basic difference between this phenomenon and that of general spectators is only that of decorum.

Keeping a scrapbook of media coverage is always a positive step to encourage volunteers (as well as paid staff). Negative articles, of course, should receive different distribution, since they often offer material for lessons to be learnt – either about working, or at least about how to handle the media.

JURISDICTION

One of the main problems in dealing with volunteer organizations, particularly smaller or nascent organizations without an entrenched bureaucracy, is in regard to jurisdiction. These organizations often take upon themselves "freelance" characteristics, rendering assistance wherever possible, rather than adhering to formal geographic boundaries. Although this flexibility has its operational benefits, it does run the risk of coming into conflict with other volunteer organizations offering similar services. Sometimes this conflict takes upon itself aspects of competition.

Flexible jurisdiction is not only geographic. There are numerous examples of volunteer organizations taking upon themselves functions beyond their charter – for better or for worse.

USE OF CLERGY

In many respects members of the clergy can be regarded as a volunteer corps, at times acting as individuals and at times representing a clerical organization. The role of clergymen can be divided into distinct categories. Since these activities are "religious" in nature, the clergymen performing them must remain primarily as volunteers in countries that have separation of Church and State.

- *Prayer:* Various religions have different prayer rituals. Roman Catholics apply "Last rites." Jews are asked to repent for their sins. Moslems prefer that a believer cover the deceased and recite verses from the Koran. In all of these cases the burden falls upon the cleric present.
- *Religious law:* Certain religions have legal requirements that are of concern to clergy in times of disaster. In Jewish law there are formalized requirements to collect and bury all parts of the dead (including his spilt blood). There are also extensive guidelines for identifying the dead.[2]
- *Assistance:* It is quite common for the clergy to volunteer assistance in times of disaster. They have a natural connection with their religious faithful. Volunteerism formalizes their position with responsible operational authorities. One effective use of the volunteer clergy was at the Swissair response headquarters in Halifax, Nova Scotia (Canada) where certain religious functionaries served as a recognized volunteer conduit to collect *ante mortem* data from families.
- *Consolation:* A key role of all clergy is to find those comforting words to console people who have lost relatives. It should be remembered that a significant part of the consolation process is to listen to a mourner to hear what bothers him. The constant quotation of verses can be a defense mechanism by a clergyman to distance himself from hearing a mourner's painful message.
- *Death notification:* A natural follow-up to consolation is death notification. After all, families need consolation since they are very much aware that the missing person has probably been killed. (This is even more the case after hospital admittance records do not show the name of a particular missing person.) The clergy, again, can only assist responsible government authorities; death notification is a legal process, the responsibility for which cannot be ignored.
- *Burial:* In most cultures the clergy is associated with running the burial (or cremation) service for the deceased.
- *Memorial:* In years following a disaster, clergy often organize memorial services to give a religious meaning or interpretation to what happened. These services are properly coordinated with authorities, since they run the risk of disorder through opening up old wounds. In the case of the Tasmania mass murder there were some complaints that the initial memorial service was designed for "politicians, dignitaries, and Heads of Churches rather than for those specifically affected by the shootings" (Smale, 2000). It should not be forgotten that these people also need a forum of expression, both for their own feelings and for their perceived job obligations.

[2] Primary classical source is Code of Jewish Law, *Even HaEzer*, Chapter 17. See Levinsohn, 1999 for a discussion of applications to modern problems.

The most effective way to handle clerical volunteers is to anticipate the process and include the clergy in the contingency planning stages of disaster response. That clergymen are not susceptible to psychological trauma is another disaster-related myth that must be dispelled. Faith does not constitute an immunity to psychological problem; instead it provides a responder with motivation and infrastructure. Those two latter factors can strengthen a responder, giving him a purpose that will hopefully carry him through the disaster.

TRAINING

Whenever possible volunteers should be given training in advance of a disaster. At the disaster site one gives specific instructions, not basic training. The training program should reflect what the volunteers have to know and what they have to do. Often one must also cope with what volunteers want to know. This is a delicate issue, since stifling curiosity can stunt the desire to volunteer. On the other hand, the training program cannot lose its work focus.

In today's mechanized and technical culture, training a volunteer can take much longer than in years past. This investment means that short-term volunteers might be more of a drain on the system than a benefit.

Training should be divided into various stages: general knowledge about disasters and the role of the volunteer, rules and responsibilities governing those persons in the relevant organization, orientation to the organization, and the job to be done. As in any teaching situation, the training program should have clearly defined objectives. The content to be taught and the method(s) of teaching should be decided upon in advance. There must also be a method of evaluating "student" performance.

A common mistake in many training programs is to explain "about" a particular job, rather than "how to perform" the job in question.

HANDLING VOLUNTEERS

Most disaster response agencies are at least semi-hierarchical, whether they be volunteer (e.g. Red Cross and Salvation Army), governmental (e.g. regional or national government), quasi-military (e.g. police) or uniformed military. This is comfortable to regular responders. There is a clear chain of command, and the authority to give orders ("work assignments" in civilian terminology) is clear. The lower a person is in the hierarchy, the more comfortable he is in receiving instructions and being told what to do. (This same attitude is true with victims; they want to be told what to do.)

Although volunteers work in this environment, they often feel exempt from supervision. This attitude must be dispelled during the course of training.

VOLUNTARY ORGANIZATIONS

Sometimes it is not individuals who offer their services to aid in a disaster, but rather an organization that incorporates volunteers and offers its organizational service. Such voluntary organizations have been known throughout the centuries. Most organizations start as an effort by volunteers to provide means and render service beyond the capabilities of individuals. In the same spirit of increasing means and service, some organizations have joined multi-organization networks.

In one major aviation disaster response, for example, a series of non-government organizations (NGOs) assisted, ranging in size from the Salvation Army to a local radio communications club (helping with communications in areas without telephone service in the pre-cellular telephone era). Some of these organizations are large and well-known; others are smaller and known either regionally or to a specialized audience. The smaller organizations tend to start by defining their own mission (not necessarily coordinated with anyone else) (Howard, 1999) and tend to operate on the fringes of statutory service providers (Poulton, 1988).

Although voluntary organizations are not officially affiliated with a government, many receive government funding of one sort or another. Sometimes this funding is sometimes a source of instability, since levels vary from year to year, precluding organized planning.

One sees volunteer organizations in response stages of a disaster, since it is in that stage that humanitarian aid is most needed. It is ironic that there is much less donor activity in mitigation, particularly related to assistance in designing programs and funding implementation. In specific terms, finances and assistance are available for the hospitalization of victims of an air crash in a developing country, but not many are willing to help authorities design and exercise a disaster response program (that can help save lives!).

A variation on the theme of voluntary organizations is the volunteer police unit. As a guide to motivation, in the call-up of volunteer dentists to assist in odontological identifications following the 22 March 1992 crash of USAir, all 22 dentists called reported for volunteer duty.

PROVIDING INFORMATION

In many disaster situations stepping forward to provide information is an act no less of volunteerism than those who step forward to work. This is particularly true when there is no established information collection network. Obviously, offering information to help find missing family members hardly fits this category; however, offering information to responders on the possible location of non-family community members, who might be trapped, does fit the definition, since it often involves hours on scene rather than a one-time declaration of facts.

All information received from voluntary sources should be examined for validity before the decision is made to invest time and resources, and go into action. There are also times when contradictory information is received (Stanton *et al.*, 2000).

VOLUNTEERISM SUBCULTURE

It is no exaggeration to describe a disaster volunteer subculture. In Israel the Civil Guard disaster volunteer unit has very much taken upon itself traits of a definite subculture, operating within the greater circle of general Israeli social behavior. The volunteers virtually all identify themselves with various segments of the ultra-religious community. On the one hand they are typical members of that community. On the other hand, they have to varying degrees a subcultural strain of specific-application volunteerism.

The Israel Police Civil Guard has general guidelines for volunteer manpower and the equipment those volunteers require. Particularly amongst the disaster volunteer leadership a subculture has developed and taken upon itself almost main-cultural proportions, transforming volunteerism from a several-hour-a-week effort into an almost full-time preoccupation. The volunteerism comes with attire (distinctive vests fulfilling the same psychological function as more standard uniforms), communications, and operational equipment.

Inter-leader communication is unusually frequent. A non-profit organization has been developed to absorb volunteer efforts that cannot be accommodated within the Police Civil Guard. Equipment guidelines provided by the Israel Police Civil Guard have been long surpassed, as the volunteers have pressed their subculture requirements on the overall Israeli "system." In every sense a subculture has been developed. It is a subculture of preparation, standby, communications, sirens, and high-profile media interest. It is a subculture that fulfills psychologically generated and religiously supported needs. But it is clearly its own subculture.

Certain Salvation Army and Red Cross units show many similar subculture traits of uniforms, volunteerism slang, and what in some cases can be coined a "disaster volunteerism way of life."

CONCLUSION

Running a volunteer program is not a simple endeavor, however volunteers can be of clear benefit in a disaster response if their energies are properly channelled. The welfare of volunteers should be taken into consideration. They are susceptible to dangers (e.g. medical, psychological) no less (if not more) than paid responders.

CHAPTER 22

PSYCHOLOGICAL TRAUMA AND MENTAL HEALTH

"If you don't believe your life matters, then it doesn't."

INTRODUCTION

The initial impact of disaster on mental health is likely to be substantial. Even though the reaction is very individual, upwards of 70 per cent of those who experience traumatic events may be expected to react acutely. Most, nevertheless, are able to sufficiently overcome their initial reaction on their own so that they eventually resume everyday functioning. This is true even though the emotional scars remain and may return to trouble them because of some future trigger. A small percentage of the victims of disaster would definitely benefit from professional mental health attention.

PSYCHOLOGICAL TRAUMA

Psychological trauma is an issue that can be foreseen, hence dealing with it can be part of an overall disaster response plan. A clear distinction must be drawn between general *psychological treatment* and *crisis intervention*. It is only the latter that fits into disaster planning.

Mental health professionals tend to approach traumatic stress from differing perspectives. Unless proven otherwise, crisis interveners presume that victims are ordinary persons suddenly overwhelmed by traumatic events in their surroundings that prove too much to handle alone. Their goal usually revolves around returning to the victims their capacity to cope.

On the other hand, many psychiatrists and clinical psychologists tend to see the emotional victims of disaster as psychiatric cases that need short-term treatment. To them, the external event is at best a trigger, setting in motion a response rooted in a flawed personality. The second approach calls for finding the flaw in the victim's personality and correcting it, if possible (Granot, 1996a).

Tumelty (1990) identifies two categories of disaster victims who are often in need of crisis intervention counseling:

- Bereaved – who search to understand what happened, and to redefine their relationship with the deceased.
- Survivors – who want to understand how and why they were saved, and how to deal with their feelings.

To these victims one must add the responders, although they tend to be treated (or at least should be treated) within their own employing agencies. It is inappropriate and problematic to mix responders with victims for intervention both because of different problems, if not also because of different perspectives (disaster participant, not disaster victim).

Even all bereaved and survivors cannot necessarily be treated by crisis intervention specialists. There are those who decline treatment. There are others who live far from the site of the disaster; they are out of jurisdiction for the main corps of social workers or psychologists, hence they often do not receive ongoing intervention, except in cases of extreme reactions.

The greater the number of victims the more wasteful is the attempt to inappropriately apply intensive therapy to the many whose normal reaction is momentarily troubled. A variety of interventions ranging from community education to mutual support, group counseling, and individual therapy can help victims varyingly affected. Intensive therapy should be reserved for those with severe emotional disabilities related to the trauma of a disaster (Granot, 1996a).

It is advisable to keep track of intervention offered to families, so that families are not approached by more than one source. Information about this kind of contact with families is, of course, personal and sensitive. It, like much other information relating to people involved in a disaster, must be safeguarded against unauthorized disclosure.

If the questions about disaster and crisis intervention are relatively standard (inquiries about dreams, listlessness, memories, etc.), the answers certainly are not. Each person has his own personality and "psychological baggage." Although there are general rules of treatment, each person must be treated as the individual that he or she is.

An accepted method of dealing with psychological reactions of responders to disasters is to have group sessions in which people who played a role in the disaster response can vent their feelings. Experience has shown that large meetings with survivors and victims' families run the risk of becoming verbal lynchings of those thought to be responsible for the disaster. Many intervention experts, therefore, prefer quieter individual or single-family sessions.

Another accepted preliminary method of reaching out to those who might need help is to issue a victims' or bereaved families' newsletter. This offers the benefits of providing factual information and general education, as well as giving an outlet for feelings and the possibility of contacting others directly involved in the disaster. Survivors are concerned about each other, particularly in disasters where there was a pre-incident relationship (e.g. a pre-existing work or social group). This relationship also extends to concern for family members of those who did not survive. A newsletter can be an outlet to show concern in this area as well. That newsletter, however, carries the disadvantages of a certain amount of depersonalization and formalizing splinter groups (as in the case of the Lockerbie family groups, each "faction" eventually developing its own newsletter). Newsletters are also more effective with those people who are better educated and can more effectively relate and react/respond to written material.

PRINCIPLES

There are a number of general principles that govern reactions to disaster and crisis intervention.

- *Principle:* It is totally normal to have strong psychological reactions to extreme events, and teaching this principle is a key element in preparing disaster workers.
  - *Application:* Disasters *are* extreme events, hence it is totally normal to have strong psychological reactions to disasters.
- *Principle:* It is totally normal to have negative reactions to negative events.
  - *Application:* Disasters are negative events (with its scenes of destruction, tragedy and injury/death), hence it is totally normal to have negative reactions to disasters.
- *Conclusion:* Although this message is quite logical, it is all too often forgotten by disaster responders, who subconsciously view psychological reaction as weakness, either in their own eyes or in the eyes of others.
  - *Sub-Conclusion:* Many people do not need a program of outside assistance to cope with a disaster. Others welcome non-intensive support. There are, however, those relatively few who do require an ongoing psychological intervention and treatment program.
  - *Sub-Conclusion:* People develop psychologically as they develop physically. A post-disaster psychological intervention program appropriate for one age group is not necessarily appropriate for another age group.
  - *Sub-Conclusion:* Outsiders can usually take care of adults. Parents and siblings are most often the best to take care of children.

PATTERN OF RESPONDERS' PSYCHOLOGICAL REACTIONS

The psychological reactions of disaster responders can work along the following patterns, stopping at any point:

	→ towards burnout
Incident → reaction	→ towards adjustment and non-problematic normalcy
	→ psychological problem → psychological disorder

How prominent is psychological reaction to disaster? These are some examples. In this case there is no real difference in transportation and non-transportation incidents. Human reaction is very much the same.

- Wilkinson (1983) noted that following the skywalk collapse in the Kansas City Hyatt Regency Hotel, all uninjured people and disaster responders had some kind of psychiatric/psychological symptom. (One wonders what person would not react to such sights.)
- Frederick (n.d.) estimated that in air crashes 75 per cent of people involved show at least some "disaster syndrome" symptoms.
- In the year after the mass murder in Tasmania in 1996 there was an 80 per cent turnover in the working staff at the Port Aurthur Historical Site, scene of the tragedy; this compares with 10 per cent in an average year (Smale, 2000). Those symptoms can be divided into clinical (requiring treatment) and sub-clinical (that psychological "baggage" that we all carry around, or in other words part of our life experience/memory).

This chapter will discuss the (1) reaction process and its stages, then (2) stress as the basic component of a reaction. It will examine (3) burnout, (4) psychological problems, then (5) psychological disorders.

STAGES OF REACTIONS

One model (Farberow and Frederick, 1978) postulates that there are four phases of a victim's psychological reaction to a disaster:

1. *Heroism.* During impact and immediate response those involved muster strength and render aid to others involved.
2. *Honeymoon.* During a period from one week to several months after the disaster victims tend to cope, being thankful that they survived.
3. *Disillusionment.* From two months to two years after the disaster, disillusionment sets in as social and government support wanes.

4. *Reconstruction.* Simply put, this is the stage of "putting the pieces back together" – both physically and psychologically.

Responders experience similar stages, although applications are slightly different. Heroism is being the saving or rescuing hero. Honeymoon is feeling satisfied from heroic performance. Disillusionment is the internalization of what happened and coping with stress from the incident. Reconstruction is the same as for the victim.

There are, of course, some responders who skip steps. A responder suffering from an acute psychological problem might immediately jump to disillusionment.

STRESS

Walter Cannon (1914) first formalized stress as a physio-psychological reaction with biological ramifications; his work was refined not many years later by Hans Selye (1979)[1] who postulated that stress could cause illness.

[1] The reference is to a summary volume of Selye's works, and not to the original publications.

There are many definitions to stress and stressors. Let it suffice to say that stressors are those stimuli with which a particular person must cope and/or accept. The confrontation with stressors is not always easy. The range of difficulty can be from virtually minimal to very substantial. The product of stressors at work is stress.

Stress can be positive (eustress) or negative (distress); very often the difference between the two is a subconscious or cognitive choice made by the person effected. Figley (1995) classifies stressors as biological, social, and psychological.

There is essentially no limit to the stressors that influence a person and combine in various proportions to influence reaction to a critical incident. On a macro level there are work–home/family–leisure–health conflicts as expressed in the time and effort dedicated by a person to each category. Most professional literature ignores leisure and health, and concentrates on work–home/family, both as individual and inter-related subjects.

Concerning the job environment and routine disaster-related tasks, there are work stressors involving goals definition, quantity load, work quality, physical environment, and customer/management satisfaction. All of these components help determine a worker's job satisfaction, which in turn is a major determinant in setting stress reaction to an incident. The more satisfied a worker is with his job, the stronger his psychological defenses are to a disaster.

Perrier and Toner (1984) analyzed stressors in the police environment. Virtually all of these stressors have direct application to the police response to disaster. One innovation of the study is isolating dissatisfaction with distorted media reporting as a stressor on policemen.

In routine functions it is often possible to control or even avoid stressors. For example, training in time management can be an effective tool in controlling workload problems. This is difficult in the initial stages of disaster response, when there is an excess of immediate/important tasks (Covey *et al.*, 1994). In later stages of the response there certainly is room for an evaluation of tasks, and deciding which tasks: (a) you should do, (b) someone else should do, (c) really need not be done at all, (d) can wait until after the disaster is over (particularly routine work over which the disaster response is superimposed).

Biological stressors include food components (e.g. caffeine, amphetamines, nicotine) that stimulate stress. It is theoretically recommended to exclude or at least restrict these items (reducing arousal) in the food and drink (and cigarettes/cigars) given to workers during an emergency response. It should be remembered that these items ingested before a sudden incident without forewarning may still be influencing the body's biological reactions.[2]

[2] There are those who recognize but minimize the effect of diet on stress and contend that the level of influence does not justify alteration of diet during a disaster response.

Beaton and Murphy (1995) talk about tasks (e.g. emergency telephone operator) that routinely "absorb" stress; it is simply in the nature of the job. Enlightened management can reduce stress, but a certain amount of stress (which should be released through planned programs) is inherent in the employment.

There is a definite connection between emotional stress and physical illness. This has been documented in DSM-III-R (APA, 1987).

There are a number of role conflicts which cause stress for disaster responders. Examples are:

- Survival guilt
- Achieve vs surrender goals
- Save oneself vs remove oneself from danger
- Function under stress vs seek treatment

BURNOUT

This is a term for which sociological/work forum application was first popularized by Freudenberger (1974), based upon the physics sense of the word, and first used in 1940. Burnout is a gradual process leading to feelings of work exhaustion and cynicism, with physical, emotional, behavioral, workplace and interpersonal overtones (Kahill, 1988). It is not to be confused with PTSD (post-traumatic stress disorder), which can be cumulative (Beaton and Murphy, 1995) in build-up, but more sudden in its actual manifestation.

Burnout is contagious. If one person starts speaking apathetically and with cynical overtones or a disdainful attitude, there is a danger that listeners will pick up the cue and gravitate towards a similar attitude.

PTSD can be a factor contributing to burnout. Burnout can also contribute to PTSD, primarily in lowering the ability of a person to cope psychologically. In many professions there is a typical "working lifetime" before expected burnout; exposure to disaster-related stress can hasten burnout and shorten working careers (Beaton and Murphy, 1995) without reaching clinical levels of PTSD or STSD (secondary traumatic stress disorder). The Maslach Burnout Inventory (MBI) is often used to measure behavior indicative of burnout (Maslach and Jackson, 1981a, 1981b).

Both proactive (preventative) and reactive (treatment) responses to burnout are recommended. These solutions include educational briefing, periodic assignment changes, and general management measures to lift morale. One of the biggest contributing factors to burnout is the nature of the job assignment, and disaster response rates higher than many more routine assignments.

"You're not responsible for what happened, just for your reactions to it."

STRESS DISORDER

Post traumatic stress disorder (PTSD) was first defined with standardized symptomatic descriptions in DSM-III (APA, 1980), then amplified and refined in DSM-III-R (APA, 1987) and DSM-IV (APA, 1994). PTSD has become a general term to encompass a large number of different and/or interrelated problems dealing with psychological reactions and problems experienced following events with major trauma.

Just as civil protection can be traced to military beginnings, much of the professional literature concerning PTSD has developed around battle fatigue and scenes of war. The Second World War saw the first beginnings of the scientific study of psychological reaction to the battlefront. Numerous studies were then conducted using data and experiences from the Korean War, from Vietnam, and from the Gulf War. These studies are of academic interest, and they are outside the purview of this book.

In the late 1980s and early 1990s another set of terms became popular. PTSD became primary traumatic stress, to which were added secondary traumatic stress(or) (STS) and secondary traumatic stress disorder (STSD). A person directly involved in a traumatic incident is susceptible to PTSD; a person not involved in the actual incident, but potentially influenced by a primary victim, is susceptible to STSD.

There are numerous symptoms that are commonly grouped under PTSD. They tend to be normal processes taken to an extreme. Mitchell and Everley (1996) group symptoms as cognitive (e.g. confused thoughts, inability to decide), physical (e.g. sweating, dizziness, high blood pressure), emotional (e.g. anger, depression, grief), and behavioral (e.g. increased/decreased food intake, withdrawal).

For example, in the behavioral area depersonalization is a normal defense mechanism. When taken to the extreme it is characterized by lack of feelings, psychological numbness, problematic interpersonal relations, and inability to concentrate. Everyone has intrusive thoughts, or recollections from previous experiences. It is quite normal to see a particular model car, then to recall an old friend not seen for years, for the simple reason that his father had owned a similar vehicle. We are constantly dealing with this type of thought association. The difference is that intrusive thoughts are associated on a negative basis; they tend to be recurrent and bothersome.

PSYCHOLOGICAL VICTIMS

There are numerous typologies to classify PTSD victims. Those victims of most concern in disaster response operations are:

- *Potential primary stress victims*
  - people involved in a disaster;
  - disaster eyewitnesses.

- *Potential secondary stress victims*
  - the people living with that person (usually family members or "significant other");
  - other family members of responders;
  - on-scene disaster responders;
  - off-scene disaster responders (e.g. hospital workers).

Each of these people can have a psychological reaction of varying severity. Reaction severity is not absolutely dependent upon proximity to the event. There are tertiary[3] and further removed "victims" (extending to such groups as television viewers); however these people do not fall into the category of primary disaster victims, nor are they typically "disaster responders."

Disaster response is not a defined-skill activity similar to a factory assembly line. An individual brings to it the wealth of his life experience. Positive incidents in the past help; negative incidents hinder. It is only hoped that the responder walks away from an incident with more satisfaction than scars. Perhaps there can be no greater success in disaster response than doing more good than harm, helping more than hurting. After a response it is sometimes the helper who needs help.

The above list is a good analytical tool, but following the Beverly Hills Supper

[3] Tertiary victims (those who had no contact with primary victims, such as the general public or more specifically television viewers) are generally not disaster workers. A possible exception would be off-scene disaster planners (who feel a responsibility for response failures). One cannot deny that disaster can have an effect on the public. This was seen in Tasmania after the 5 January 1975 disaster involving the Tasman Bridge (McGee, 2001), and in the decline in bus ridership in Israel after a mid-1990s wave of terrorist bus bombings.

Club fire a far more simplified order of priorities was established: survivors, bereaved, responders. Each group had its own unique problems, and each group had its own point of initial treatment.

Pearce (1985) reports that PTSD is stronger in deliberate and human-created disasters than in natural events. Fleming and Baum (1985–86) reinforce this same idea and point to technological catastrophes (i.e. disasters caused by the failure of human-designed systems) as causing longer-term stress. Eyre (1999) cites anniversary dates of a disaster as a delayed trigger for the onset of PTSD.

EMOTIONAL CONNECTION AND SECONDARY VICTIMS

There is a psychological process of identifying with a person in crisis. There is no problem in feeling for a victim, and there are those who suggest that empathy is a desired feeling. A mental health worker, however, should always be careful to maintain a professional relationship with victims, and not allow personal feelings to obscure true goals. Following the Piper Alpha explosions in the North Sea, there were regular debriefing sessions for the therapists and their supervisors who were involved in dealing with the psychological problems of primary victims. Interviews were also conducted with the treating medical staff, including nurses (Alexander, 1991).

It is a natural reaction to find some kind of sympathetic connection with disaster victims. After all, the status of victim is most often without direct fault on the part of the person involved. This phenomenon of sympathetic connection is also to be found amongst therapists, who try to sort out the psychological problems of their patients. Corey (1991) talks of transference and counter-transference of emotions between therapist and the person treated.

It is all the more common to search for an emotional connection with disaster responders, since the therapist is, himself, a disaster responder.

There is a fine line between identification with a victim and empathy with him. Empathy can lead to intense connection with the victim, even to the point of taking over his complaints and fighting against those perceived as wrong-doers. Empathy can also lead to responding to the stress of a victim with your own stress. Responders must understand their role and responsibilities, and maintain a victim/responder relationship in its proper context.

IDENTIFYING PTSD

Ascertaining who is suffering from various stages of PTSD is a complex issue with different approaches. Some professionals prefer a trained mental health professional at the disaster site to screen for problems. In a fast-moving response a person will not see the entire operation, and at best he will catch only the most immediate and blatant expressions of stress (not an unimportant achievement).

After an incident people will talk about what happened. Mitchell and Everly (1996) view this debriefing – Critical Incident Stress Debriefing (CISD) in their terminology – as part of an entire treatment process (Critical Incident Stress Management, CISM) and not as an isolated activity or one-stop fix-all move.

Their advice is to conduct debriefings ranging from immediately after an incident to one to three or four days later. A same-day debriefing has its benefits, but it does not replace a later session when people have had time to consider their experience and to react to it. The motives of a debriefing session are to explain that stress is a normal reaction and to allow people to vent their feelings (Roemer and Borkovec, 1994). Acute problems can usually be identified through this venting process. Mitchell and Everly (1996) describe a seven-stage process in the debriefing session:

1. Introductory remarks – setting the tone
2. Fact phase – what happened?
3. Thought phase – what were you thinking?
4. Reaction phase – what was the worst part?
5. Symptom phase – distress signals
6. Teaching phase – provide information
7. Re-entry phase – answer lingering questions

On a very practical level these debriefings may be administratively difficult to organize given short notice, work schedules, and funding allowances for a relaxed setting. They should however, be, a standard part of the disaster response process.

The debriefings should be mandatory for everyone involved in a disaster response. The debriefings are a group activity. For bureaucratic organization responders it is recommended that the command level be dealt with separately so that individuals are not inhibited from talking freely in the group. For this reason Mitchell and Everly (1996) stress the use of a peer counselor (alternatively, peer support personnel). There must also be a firm caveat that conversations are privileged and not to be reported to others.

It is the unaffiliated volunteer who is usually left out of the debriefing process. He renders assistance at the most stressful time, before professional responders arrive at the scene to bring order, then he returns to his anonymity (even as the response continues). Since he is "unaffiliated," no one feels responsible for him (in many cases not even knowing his name).

Aviation accidents also provide a unique situation. Airlines often have only limited contact with those passengers not suffering physical injury. Once their medical status has been verified and they have been questioned regarding the accident, they travel onward (away from the difficult experience) – before enough time has passed to allow for a meaningful psychological interview. Providing these people with contact information in the event of psychological crisis highlights the acute and transforms what should be a mandatory procedure into a voluntary exercise.

Sometimes disaster response is complicated by outside factors. Following the Beverly Hills Supper Club fire, clergy were used as a psychological help in dealing with bereaved families. Viewing of bodies started at 0500 am, but soon thereafter many clergymen realized they would have to depart the scene to tend to their Sunday morning congregations with the probability that parishioners might have known some of the victims.

TREATING PTSD

A recognized handling of PTSD is to allow a person to repeat his experience (ventilation). Unless this process is properly supervised, the retelling of events can have a negative, not positive, effect in which the teller relieves his trauma each time he reviews what happened. This author saw this phenomenon when a hotel manager broke into a sweat as he recalled a hotel fire on the tenth anniversary of the event. He had been the shift manager at the time, and he still held himself to blame for the tragedy and loss of life.

It should be realized that there are various levels of psychological assistance. After a disaster most assistance is rendered by social workers or other human service workers who have undergone limited training about psychological reactions. There must be a program in hand for these workers (who are generally volunteers, with all due respect) to refer difficult problems to the professionals. Problems can be defined as irregularities in alertness, awareness, responsitivity, speech, and emotions (see Farberow and Frederick for detail).

Too many treatment programs are only cognitive. It is hard to treat psychological phenomena only on a rational basis. On the other hand, the study of emotions has a non-scientific connotation; emotion-oriented programs must overcome that stigma.

The term "treatment" is itself suggestive of a directed effort by someone who is responsible for "curing" the victim. Such a relationship, psychologist–patient, might be accepted in a medical scenario, but it is difficult in the psychological domain. People do not like to be labeled as "having psychological problems," even when those problems are quite acute. Often this reluctance is influenced by occupational, organizational or social impediments. A much less sensitive term is "counseling" or "debriefing." Linguistic name changes only help; alone they do not solve problems deeply rooted in social conventions.

Another common factor complicating treatment is the effort by some people to cast blame on those thought (sometimes irrationally) to be responsible for the incident causing the traumatic situation. Sometimes the "guilty" are the same people offering psychological assistance. Due to this situation it is best that psychologists/therapists be independent and be brought in from the outside.

In view of the perceived stigma of dealing with psychological reactions, people tend not to seek assistance openly. A proactive program is most often needed to reach those in need.

It is usually a clinical psychologist who is assigned the task of treating stress disorders. There is much to be said for this in the treatment of non-acute cases. Social workers can also be used effectively to discern disorder symptoms.

Mitchell and Bray (1990) phrase a treatment strategy as REAPER (R–recognition, E–education, A–acceptance, P–permission [to treat], E–exploration [of treatment resources], R–referral). Mitchell and Everly (1996) speak of SAFE-R in the Exploration stage: Stimulation reduction, problem Acknowledgement, Facilitation of treatment and symptom normalization, Encourage coping techniques, Restoration of function or further Referral.

COPING WITH TRAUMA

Not every problem has to be treated. Sub-clinical problems can be handled by the person himself (often with routine "coaching" that would not qualify as "treatment"). Valent (1995) identifies several strategies in the process of coping with trauma. Although quite theoretical, many of the categories (or versions thereof) have direct application in disaster scenarios:

- Rescue – the need to give
- Attachment – wanting others to protect you
- Assertiveness – striving to achieve goals
- Acceptance – adapting to a situation by changing goals
- Fight – in war situations
- Flight – evacuation
- Competition – the fight for resources (whether needed or not)
- Cooperation – affiliation with other people (e.g. victim/responder)

Attachment is a common trait, not restricted to disaster situations. In many situations people show attachment to other people and to property – and they expect government to protect those attachment links. That protection is one of the main features of many government disaster programs.

Assertiveness is the contradiction of the myth that disaster victims are helpless. They do assert themselves in carrying the burden of rescue, particularly in the immediate aftermath of catastrophe. Responders should not contradict assertiveness by conveying a message of "taught helplessness." This same assertiveness helps the rescuer, as he imposes his goals on the disaster response.

Responders experience severe problems when they are unable to accept disaster situations, particularly as they require a change or adaptation of goals. This comes into play when a person can no longer be rescued, or when a victim no longer needs assistance.

There is a distinct difference between fleeing and distancing oneself from disaster. For disaster workers, these coping mechanisms have the effect of premature withdrawal of response services. The worker who flees simply withdraws physically; the worker who distances himself withdraws psychologically. Although over-identification should be avoided, a proper medium of involvement should be sought.

Competition is a common phenomenon that often takes upon itself a self-importance that obscures real problems (hence forming a psychological defense against those problems). For the disaster worker this often means getting to the site before others (even though services are needed only in a later stage, and early arrival only places unnecessary strain on the response system). Competition also manifests itself between groups or organizations, particularly when there is a hierarchical void.

An important element to assist persons exposed to trauma is family support. For disaster victims this can be particularly problematic, since often their families are also victims. For responders, their families often need guidance to supplant family worry with family support.

SEARCHING FOR SCAPEGOATS

A common reaction of disaster victims is to search for a responsible scapegoat and blame him for the disaster's outcome. This can take the form of an assault on those very people who have been of assistance. Sometimes the assault is expressed in "cultured" terms – a lawsuit. In other cases that assault can be just what the word means – a physical attack.

COMPENSATION

There have been disaster responders who have felt the right to receive special compensation for the trauma of difficult incidents that they handled in the course of their jobs. On the one hand that trauma can be very real with potentially serious after-effects. On the other hand, if trauma is handled properly, there should be very few cases in which compensation is a legitimate factor. In any event, claims for psychological injury should be handled in the same way as any other work-related or on-the-job injury.

THOSE WHO "COULD HAVE BEEN THERE"

The statement, "I could have been there [at a disaster]," is often heard. Using aviation as an example, such a claim can be divided into several categories:

- Having reservations on a flight that eventually crashed, but having (a) changed reservations, or (b) missed the flight.
- Traveling on the previous flight using the same number.

In these cases the degree of emotional tone of the statement can indicate whether the speaker is, in fact, calling for psychological help or counseling. If the person making the statement flew the route long ago or flew more recently on another carrier, it is likely that he is looking for attention, not help. The reader can apply these examples to other types of disasters.

In all of these cases, "I could have been there" is part of the attitude of fate that subjectively accompanies acceptance of disasters. That element of fate is also a key element in the media coverage of air disasters (Garner, 1996).

TRAUMA AND THE GENERAL POPULATION

In a number of incidents there have been public reactions that have gone well beyond the usual experience of sorrow and dismay. In Israel, for example, after the wave of bus bombings during the mid-1990s, fear took on active proportions as the number of passengers traveling on buses took a serious drop. In essence, the repeated incidents had the effect of destroying the "security culture" that permits people to function without fear.

CONCLUSION

The science of dealing with stress is a fast-moving field that has undergone significant change in progress since the late 1970s. Current thinking is unanimous in concluding that all disaster response programs must include procedures to cope with psychological problems, both of responders and of victims.

CHAPTER 23

MEDIA AND INFORMATION

INTRODUCTION

The release of accurate information during a disaster is often of utmost importance in maintaining order and saving lives.

- When officially released information *is not* available, people often turn to private sources, opening the door to misinformation and rumor.
- Even when accurate information *is* available, people always seek multiple sources, informal and formal, both to act and merely to satisfy curiosity. This can be seen in listening to several radio and television news broadcasts about the same event; invariably, the stations carry the same news, if not the same pictures.[1]

[1] Levinson (1998a, 1998b) shows that newspapers also give essentially the same coverage of an event, except for tangential stories.

Often the information sought by the public regarding a disaster is by no means critical to the response. Even though the information might fall within the realm of mere curiosity, a steady flow of facts is often required.

PROVIDING INFORMATION

In the United States coordination of disaster information usually falls in good measure on the local municipality, at least in the early stages of a response. This is due primarily to the jurisdictional convention in which local government operates fire and police – two of the most prominent responders. This is not to say that the procedure is without problem, since contradictions in information even between these two sources have been known. For this reason it is essential that different agency spokesmen work together.

Information released by a spokesman is *controlled*. When this is done professionally, "controlled" should not be equated with "censored." Any good spokesman knows that the truth cannot be hidden for any extended period of time. "Control" is verifying facts, not yielding to rumor, and holding back information about personal injury until families can be properly notified. In disasters suspected to have been caused by willful criminal acts, it is also the non-publication of investigatory information needed to "solve" the case.

Verification takes time. When controlled information is slow in coming, it is a natural reporting inclination to seek uncontrolled information to fill the void, even though it is known to be less reliable (or not reliable at all).

If "controlled" information is verified facts, analogy says that "uncontrolled" information is not verified. That uncontrolled information can be extremely fascinating and emotional, but without verification it is problematic in providing a better understanding of events.

Verification takes time. When controlled information is slow in coming, it is a natural reporting inclination to seek uncontrolled information to fill the void, even though it is known to be less reliable (or not reliable at all).

A bomb was detonated in a taxi in northern Israel on 1 March 2001. Initial reporting related that *apparently* the police had been following the vehicle based upon intelligence information. Several hours after the bomb explosion the Chief of Police explained that a roadblock had been set up based upon intelligence information. The erroneous version of events was not specifically refuted in most coverage; the newer version merely replaced the older version. Many recipients were left still believing that the police were following the taxi.

CELLULAR TELEPHONES

The cellular telephone has ushered in a new era of information availability. In many cases the cellular telephone has transformed victims into reporters, or if not quite that far, then into important news sources. Witnesses to a disaster have sometimes telephoned their testimony by cellular telephone. E-mail and other informal Internet connections have also aided in proliferating "uncontrolled" information. By its very nature, uncontrolled information can be available and transmitted very rapidly.

On 11 December 1998 Thai International Flight No. 261 (Airbus A310-300) departed Bangkok (Thailand) at 1740 for a less than two-hour flight to Surat Thani, where weather was difficult with limited visibility and heavy rain. The aircraft was on its third landing attempt when it crashed into a rubber plantation three to five kilometers from the airport; 101 people died.

From an information perspective, the TG 261 flight is an example of passengers using cellular telephones to report the landing difficulties. From a flight safety perspective it seems that passengers' cellular phones interfered with navigation equipment (though not causing the accident).

MEDIA EVENTS

In Western culture disasters are media events. Their carnage and destruction inevitably arouse public interest, and as a result they encourage coverage. Even distant disasters generate extensive media attention, as long as the disaster is sudden and somewhat relevant to the news recipient. The crash of the Air France Concorde made headlines throughout Europe and North America, because television reporting brought drama into the viewers' living rooms, and news recipients felt they could identify with the disaster. Crashes in the Third World often go virtually unreported in the foreign popular press. The press coverage is lacking, hence the element of the spectator is lost. (Famine does not generate interest in developed countries, since the news recipient cannot identify with the phenomenon. There is also no fast-moving action, such as the race against time to save earthquake or air-crash victims.)

Often the complaint is heard that media make the event. The contention is that the presence of cameras and reporters heightens the atmosphere, adding a sense of excitement to the scene. If this is true, it is only so in part. Small media crews cannot transform an incident into a disaster; large media crews are present only because the incident *warrants* the coverage. Press crews can report in a tone of drama, but they cannot create what does not already exist. Media coverage is, in the long run, a commercial operation. It begins, continues, and terminates according to economic viability – the need to produce coverage that will be transmitted to the public.[2] Its internal ethical system also rejects fabrication (though not necessarily exaggeration).

At first the news recipient is generally not looking for in-depth cause analysis. He is looking for entertainment, emotionalism and human interest. He is fascinated by the crying survivor, by the traumatized witness. He is looking for a story – with a dramatic opening, a plot, a villain, victims, and a hero. Disaster provides all of these elements. It provides a dramatic story, the plot of rescue, the villain responsible, the injured and dead victims, and the heroic rescuers who are ready to sacrifice all to save human life. Only at a later stage is the news recipient looking for more than a very superficial analysis.

The media (whether newspaper, radio, television or Internet sources) are ready to provide coverage. On television and radio that coverage, at its height, is without commercials, but it draws viewers. It improves ratings. For newspapers, it sells copy. The "extra" edition is a thing of the past. The broadcast media

[2] Academic research is motivated by the value of information collected and its potential use in situation analysis, although one might cynically claim that the resultant research reports are akin to commercial products that must be marketed (to publishers, evaluation boards, etc.).

satisfy immediate news requirements. The newspaper serves as the next morning's incident recap and photographic record.

The recipient's primary concern is that he must be insulated from personal endangerment. Yes, the news recipient must be able to identify with the incident. But, he must feel that he is protected and not more than theoretically vulnerable. He "could have been" on the airplane or train – but he wasn't. It is only in the hours after the excitement of the event that the news recipient is concerned with an in-depth analysis of the details leading to cause. Eventually, he wants that analysis. It offers a reason why the disaster happened. It brings an explanation – it brings order to chaos, and again insulates the viewer: "No, it can't happen to me!"

This produces a basic paradox in the role of the media. They play both positive and negative roles in the eyes of responders. From a response perspective, the media have an important function in disseminating information and forming public opinion. Television and radio can gather and report news very quickly.[3] Radio moves in first. Television generally follows (since its equipment is more intricate). More than once response elements have first learnt about the occurrence of a disaster through media reporting before notification was received through established official channels.[4]

[3] In terrorist incidents in Israel, television transmission from smaller cities has been relatively slow and based upon telephone and other non-visual communication until camera teams can arrive from the main cities.

[4] First notification happened this way on more than one occasion to this author. In one case cellular phone notification from a bystander preceded both media and official channel notification.

Media cooperation with government is needed in spreading announcements regarding possible dangers and evacuation requests. From a government perspective it is also desirable that the media use the disaster subtly to teach the public how to behave or react in any future disasters. From a very theoretical point of view, it would be ideal if a television viewer looking at a train derailment report were to learn how to evacuate a train safely.

On the other hand, the media have their own agenda. In democratic societies they determine in great measure what information is passed to the public. As disaster reports are received in the newsroom, it is the media staff who screen information and decide what is passed on and in what form. Sometimes the selection is not totally objective. It is based upon packaging, with preference given to the tight interview, the short graphic film clip. Other times there is little screening – sometimes too little. Virtually everything is passed on to fill an information void.

The media can be caught up in their own commercial competition. There is pressure to be the "the first" with a story, to beat competitors in providing the information. That competition exists both between reporters and between news media (newspapers, television, etc.). Sometimes the rush to be first overshadows the ethical goal of total accuracy. The problem is that when mistakes are made, they become a matter of public record. Particularly in the case of newspapers, they become a permanent record. If "the truth never catches up to the lie," it also never catches up to the mistake. In many cases it is hard if not totally impossible to correct mistaken information . . . and the *impressions* made by that information.

TERRORISM

A total news blackout on terrorist attacks would be a helpful factor in denying terrorists the publicity they so badly seek, however such a step is contrary to media needs. The terrorist wants publicity. He thrives on headlines that, in his mind, carry his message to the population. News coverage also spreads the word of terror – disruption in the course of usual life unless demands (whether specific or not) are met. A news blackout is a measure that cannot be contemplated in any free society (although it is practiced in army wartime operations under the same claim of "security"). Even such phrases as "war on terrorism" cannot counter the media's desire to quench the public's thirst for disaster coverage.

INFLUENCES ON COVERAGE

News reporting is the product of a reporter's experience. Today reporters tend to specialize. There are police reporters, political reporters, and city government reporters – but there are no disaster reporters. Nor are there many reporters who specialize in regulatory agencies, where much disaster-related background material can be found (Coglianese and Howard, 1998). In most localities disasters just do not happen often enough to justify hiring special staff. The net result is that when a disaster does occur, covering the story does not go to specialists. It goes to reporters usually covering other beats. And those covering the story will generally view the event from their regular working perspective. If they have previous disaster experience (as with many reporters in cities such as Jerusalem where terrorist incidents are all too common), previous incidents do promote understanding of disaster phenomena. That experience, however, cannot be considered a replacement for learnt expertise.

Different media sources have different coverage policies, each projecting a particular company policy. To cite one non-disaster example, on 26 December 2000 a disgruntled employee in an office in Wakefield, Massachusetts (USA) killed seven fellow employees. Local Boston television stations covered the event very differently. WHDH-TV (Channel 7) had continuous coverage; WBZ-TV (Channel 4) interrupted programs with long reports; WCVB-TV (Channel 5) opted for only limited incident reporting. Fox news assigned the label "mass murder,"[5] reminiscent of disaster operations. This difference in coverage is muted when there is pool coverage, with different competing television stations feeding off the same video link. Although there is a difference in the verbal content, the video dominates the coverage.

[5] Just as with "disaster," the term "mass murder" is subjective. There is no fast rule according to which the killing of "several" victims can be termed "mass murder."

Not only do the media decide to call an incident a "*mass* murder" or a "disaster"; they also assign relative importance to an incident. On the evening of 26 December 2000 the Wakefield murders were the lead item in the national news of the three largest US television networks (*Boston Globe*, 27 December 2000). The net result is that a major disaster in a "quiet" corner of the world can be virtually ignored, whilst a small event in a high news-profile city can receive extensive coverage.[6]

[6] Car/bus bombings in Israel provide an example of news distortion. When bombs exploded in Hadera and Netanya, television coverage was delayed until a film crew arrived. Jerusalem incidents receive more extensive and more immediate coverage, not only because of importance, but also due to the readily available television crews and presence of the foreign press.

Sometimes entire events are blown out of proportion; at other times the various stages or features of the event are given inappropriate coverage with regard to their relative importance. In a disaster it is common that the media have access to only part of the incident. Cameras cannot photograph "everything," nor can "everyone" be interviewed. Sometimes the relative importance of various aspects of a disaster and its response are distorted by media accessibility, under the mistaken reasoning that if it was not covered, it could not have been important.

In a disaster there is a certain "honeymoon" period during which the media relay information as it is received from government sources. After all, in the initial stages of a disaster there is a dearth of information available, so any tidbit is welcome. (That can also be problematic. As the media search for information, particularly to fill television and radio time, there is the danger that they will use unverified stories that do not represent a factual picture.)

CRITICAL EXAMINATION

Perry and Lindell (1996) found that the media tended to sustain popular disaster myths, rather than providing accurate information. The basic reason is that they drew information from "popular knowledge," and not from professional sources. (Another known phenomenon according to some is that the media at times even create popular myths.) This is particularly true in the "honeymoon" stage of reporting.

Only after there is an increase in information available, and after psychological calm is restored (sometimes well after the disaster), do the media again become a *critical press* that evaluates information and sources, and evaluates response performance.[7]

[7] Private conversations with British and Scottish reporters confirmed that for more than a week after the PA 103 crash it was felt inappropriate that they should act as a critical press. This is certainly not typical of Western press. The reaction is probably best explained in terms of psychological reaction.

Examining a disaster from the political, police and local government perspectives can be an enlightening experience, but the situation often yields the unfortunate result that the perspective of a disaster response professional with a broad perspective is missing. Most reporters are also unfamiliar with the technical causes of a disaster (e.g. basic aerospace dynamics, or the chemistry of explosions).

Differentiation must be made between a terrorist or criminal-related disaster in which the release of incident details might hamper police operations or encourage future terrorist operations, and unintentional disasters where there is a desire for information to reach the public through the media. Each type of incident engenders a different relationship with the media. In both types of event, however, the public must have details about rerouted traffic, emergency telephone numbers, requests for information concerning missing persons, etc.

In any case the basic rule of the media is to gather and then disseminate information. The general recommendation to disaster responding agencies is to provide information to the media, both informally and in press conferences. Be proactive, not reactive. Some press officers have worked under the theory that it is better to give information rather than let reporters find their own stories. This is a fallacy, since a good reporter will always turn to his own resources. If, however, official information is released, it will give a factual basis for subsequent reporting. The lack of technical knowledge regarding the cause of a disaster can be rectified with interviews, both for background and for quotation, with professionals from government supervisory agencies and equipment manufacturers.

Most government agencies and many large companies have press officers and public relations experts. Although these press officers do take a good part of the burden off operational staff regarding the media, there still is a strong desire (and justification) to interview senior personnel from responding agencies. These interviews should be conducted in coordination with the press officer.

Information released must, obviously, be absolutely accurate; otherwise it serves no purpose. Later, erroneous reporting can even serve as one of the bases to criticize the disaster response. A problem frequently encountered is that accurate information is usually not available in the early stages of a disaster response, when the press is most often in need of facts. As a result initial reporting is rarely very accurate or complete. Following the Kegworth air crash in England, the aircraft was mistakenly identified as a DC-9, rather than a Boeing 737-400, an error which kept occurring well into the incident (Smith, 1992); such an error can carry with it response implications.

ERRORS OF FACT

When official information is not available, and less reliable sources are used, rumors can start (and sometimes not be stopped even when accurate information becomes available). Coupled with a natural psychological tendency to take worst case estimates, press headlines sometimes make a disaster much worse than its reality. In the Loma Prieta earthquake (7 October 1989), for example, 68 people were killed in a ten county radius (42 on the Cypress Viaduct) (Stark,

1990), yet in early reporting various newspapers cited very different statistics (Rogers and Berndt, 1992):

San Jose Mercury News	150
Marin Independent Journal	200+
USA Today	250+
New York Times	270

As Rogers and Berndt (1992) relate, initial media coverage was centered around San Francisco–Oakland because of its proximity to reporters already stationed in the area, rather than further south in Santa Cruz (much closer to the epicenter) where a shopping mall collapsed. A contributing factor was that travel to Santa Cruz was difficult due to roadway damage.

In other cases initial numbers for injured and fatalities have been kept artificially low, "hoping for the best" (a psychological ploy to reject what is happening).

Broadcast media work much more quickly than the printed media. Newspaper reporting is dependent upon press deadlines, always molding the reporting of a developing story in such a fashion that it not be perceived as stale news once the newspaper is printed. For this reason pictures are a key part of the disaster coverage.[8]

[8] Some reporting and use of pictures can be in bad taste. Local television evening news reports carried pictures of a deceased woman killed on a terrorist-related bus crash in Israel on 6 July 1989. It is through this "notification" the woman's sister learned of the death and came to the morgue to look for the body. She was so certain of events that police waived fill-in of most of the ante mortem information reporting form.

STAGES IN REPORTING

In a developing story there are three stages of information interest: (1) what happened? (2) to whom did it happen? and (3) who is responsible? In the broadcast media coverage follows each of the stages. In newspapers press schedules can sometimes influence the omission of a stage.

WHAT HAPPENED?

In most disasters, the initial stages of reporting reflect "worst case" fears, and initial estimations are scaled down as more information becomes available. Part of this process is due to the shift in reporting and response concern from immediate emergency activities to basic information gathering. This is understandable, since in initial stages emergency services are usually involved in life-saving activities. Only after medical triage and hospital transfer are organized can a general picture of the situation be obtained.

Lack of information has a direct effect on the public. In the 21 August 1995 explosion involving the No. 9 and 26 buses in Jerusalem, initial broadcast news bulletins failed to mention the bus numbers involved. In the second stage the

No. 9 bus was reported as being No. 4. Only after about an hour were the two bus numbers accurately reported.[9] This error in the number might initially appear not to be understandable, but a moment of thought will yield the reality that the digitalized numbers on the buses were damaged, and in the immediate emergency response there were many questions more important than questioning injured survivors about bus numbers. Although the number might seem to be an insignificant detail against the background of life-saving activity, that bus number was critical information for those radio listeners trying to determine if their loved ones were possibly involved.

[9] There was a No. 4 bus in the area, a fact that complicated reporting.

Reporters always seek eyewitness descriptions of tragedy, since the testimony tends to be extremely emotional. For the same reason the same testimony should be taken with a certain amount of skepticism, since it can be error-prone. Very often emotional testimony is repeatedly broadcast on television and radio. It should be remembered that these broadcasts are repeated for emotional impact and viewer curiosity, rather than any real "need to know."[10]

[10] One must differentiate "right to know," "want to know," and "need to know." These basic principles often become confused during press coverage.

At a certain stage it becomes relatively clear what happened in the disaster. In the case of a plane crash, for example, the location of the incident and the number of passengers become known, as well as a reasonably accurate estimate of fatalities.

TO WHOM DID IT HAPPEN?

Once it becomes clear what has happened, focus usually shifts to the victims. Unless an extremely important issue arises, such as a military action downing a civilian plane, people start to concentrate on the victims. Who were they? (as though a person in Iowa really thinks he might know passengers from New York). The public is generally curious, and again the media tries to satisfy that curiosity. Although there are instances in which people recognize victims in far-off places, the conveying of news with a practical application is generally not the point, except in media local to the disaster or local to a number of victims.

There are different traditions in the world press. Some authorities contend that personalization is a trend in the English-speaking press (United States, Canada, United Kingdom). This is certainly the case in Israel's Hebrew press, particularly given the small size of the country and the higher probability that "someone knows someone else." In very broad generalization, the French press tends to be much more aloof and philosophical with less emphasis on the personal details.[11]

[11] James Lederman, personal discussion.

Names of disaster victims are generally not released until a reasonable effort has been made to notify families. While there is an obvious argument that families should be contacted to tell them privately that a loved one was involved in a disaster, there is a counter-argument that names should be released quickly so that the many more families of relatives not involved will feel relieved. In the aviation industry it is generally accepted that passenger lists be kept private; a noted exception was the Singapore Airlines decision to publish the passenger list of their crashed flight SQ 006 on an Internet site. A company spokesman explained, "SIA tried to contact all the affected families before the passenger list was released, but some were not contactable. Therefore, the list was released without notifying everyone concerned . . ."

A key element in the communications dynamics is the modern reality of cellular telephones through which survivors contact relatives (and the press) without waiting for airline-provided telephones.

WHO IS RESPONSIBLE?

It is a common public tendency to place blame, to search for culpability. It is not enough to understand a disaster. Many people feel they must find who is "at fault."[12] They must find the villain. That is part of the drama of disaster. Someone must be held at fault. In the case of natural disasters one can still place blame. One cannot fault nature for causing a hurricane or earthquake, but even then there is sometimes a villain, even if it is a municipal clerk who "should have, but didn't", or a regulatory agency that did not enforce its own requirements. Generally, the more prominent the villain, the more satisfied is the public.

[12] Just as the public searches for the culpability of others, responders often look for their own legitimization, seeking to justify their efforts as necessary, if not critical, to the response (even though they might have been quite tangential). These responders often exert pressures on the media to highlight their roles.

HANDLING THE MEDIA

Proper media relations do not begin with a disaster. Coverage during a disaster should be a normal extension of an ongoing relationship of periodic background briefings and routine interaction. If ongoing relationships are continued correctly, reporters should acquire strong and trusted connections for getting the news from official sources and reporting it accurately. In theory, those reporters should also have the background required to understand events. This, however, is truly quite theoretical. In practice the press has essentially no interest in disaster planning (Quarantelli, 1996), following the public's apathy. A corollary is that quite often the media, themselves, have no disaster plans beyond basic telephone numbers.

It is obvious that this continued relationship is not always possible. In many disasters reporters from outside the area descending on the disaster will outnumber local reporters by far. If an ongoing relationship is established with the press, it is only with local reporters. Thus, agencies responding to a disaster are often in a situation where work cooperation built up carefully over time involves only a small part of the press covering an incident. Even if there is an assistance arrangement with another press officer with wider geographic coverage (such as at the state or federal level, in US terminology), the press will arrive well before the government representative.

There are four guiding rules that govern the local/outsider news relationship:

1. Outsiders will go away after a disaster. Local reporters remain.
2. Local reporters tend to have best access to local news sources. Much disaster reporting by outside news services is, in fact, picked up from local reporters (Quarantelli, 1996).
3. Outsiders often shape the national or international image of a disaster and its response.
4. Local reports have a strong influence on the local area and on post-disaster public reaction and possible corrective legislation.

It is not a matter of giving preference to one type of reporter or another. Based upon the second point it should be very clear that outside (in American terms: regional, national, international) and local reporters cover news from different perspectives. They need different input to write their stories.

A proper disaster response relationship with the media is for press officers to have a basic proactive policy that can be followed throughout the response period, rather than passively reacting to media initiatives. Such a policy means the planned release of information through media bulletins, in-person interviews[13] and press conferences. It also means projecting an image of professionalism and competence that are important in reassuring the public.

[13] One study has shown that dealing with the media can take up to 40 per cent of a senior investigator's time during the initial stages of a serious crime investigation. There is no reason to assume that the media would monopolize less time in disaster response (Feist, n.d.).

One of the best ways to stop the spread of rumors is to supply reliable official statements properly presented to the public. Rumors are a constant problem. On 28 March 1979 a partial meltdown released radioactive material near Middletown, Pennsylvania (USA) at Three-Mile Island. Houts and Goldhaber (1981) reported that during the first year after the incident half of the population believed rumors about miscarriages, birth defects and health problems. People also reported perceived, but non-existent, symptoms. The El Al crash in Amsterdam also brought a large number of health complaints, particularly *after* public discussion of possible hazardous material reportedly carried on board.

PRESS OFFICERS/SPOKESMEN

Government offices and most larger companies (particularly airlines and railroad companies) have a designated spokesman for routine work. When he functions properly, the media is aware of his name, background, and contact numbers (including off-hours). If a company without a spokesman is involved in a disaster (as has happened to certain maritime companies), such a person should be appointed immediately (albeit untrained, but at least to serve as media-issues coordinator and focal point).

It is also imperative that company employees at all levels understand who is the company spokesman and that *only he* speaks with the media. This is done to insure a single line of explanation to the media.

In particular radio and television search for a local spokesman to interview following a disaster. If, for example, the headquarters of an airline involved in an accident is located in a faraway city, it is often preferable (at least from a media perspective) to have someone in the affected city function as spokesman.

The spokesman function needs back-up personnel, since accidents can happen at any hour. The presence of international press can also mean press deadlines that do not necessarily conform to standard working hours at the disaster scene. That back-up also has to cover the period when the primary spokesman is away on business assignment or simply on vacation in a distant city.

As macabre as it might sound, basic disaster announcements should be prepared as part of contingency planning, so that blank lines can be filled in with details, and announcements be released. This is desirable, since there is considerable pressure when a disaster has occurred. Preparation in advance allows for careful choosing of message wording and content.

It should not be assumed that press officers can properly represent their employers to the media during disaster without special training. Any media policy must start with training, both in how to deliver a message and what message to deliver. This includes people designated for press functions in times of disaster when extra staff are required.

In an air crash a significant amount of media coverage will involve:

- the *safety record* of the airline, specific aircraft, and type of aircraft involved;
- the *background* on the pilot (and to a much lesser extent other members of the cockpit crew), including training and flight experience;
- *general stories* about aviation and accidents.

It would be prudent for an airline to keep such information on file, so that it can be released rapidly as required. That use need not be due to an accident with the airline's own aircraft. When a particular model of airplane crashes, sometimes basic questions are raised even about planes of that model operated by other carriers. For example, accidents with Airbuses generated general media coverage of Airbuses, sometimes with a list of carriers with that plane in their fleet.

When there is an accident (either on the airfield or with the airfield as the destination), an airport can expect media coverage about its prior safety record, including stories recalling accidents that occurred before current institutional memory.

Coverage of train accidents follows very much the same pattern as with airplanes.

It is not only those directly involved who need press representation. This also includes secondary sources such as insurance companies, since they are often contacted by the press for reactions to accidents.

Wood (2000) makes the point that the worst response that a spokesman can make is, "No comment." In media terminology that statement can be conveyed to the public as, "The press spokesman refused to comment," or, "He could find no answer to the question." Wood also adds that there are no "off-the-record" comments; it should be assumed that everything said will be reported.

It is important to dress appropriately for a television or photographed press interview. If in an office, the background should be chosen carefully to create the proper impression. As Wood points out, at a disaster site the public is not interested in seeing a company executive dressed in a "tie and blue pinstripe suit." That attire sends a non-verbal message of protected position and non-involvement. No tie and rolled-up sleeves are more suggestive of involvement in the incident. Polarized lenses or sunglasses should also be avoided, since they prevent eye contact (the psychological reaction of the viewer is an inclination towards distrust).

These rules apply to the press officer. Image is all-important. The policy toward a CEO might well dictate a certain distance, emphasizing managerial control as opposed to personal involvement in response operations.

SENSITIVE SUBJECTS

A responsible press officer will choose appropriate language when declining questions regarding certain subjects. It is a common courtesy not to report the names of the dead until relatives have been notified, a process that can take more than just an hour or two. This ethical limitation also extends to showing pictures in which the dead can be identified. When family members of deceased persons have not been notified and names cannot be released, this should be explained to the press. This is particularly true when more than a short period of time has elapsed, perhaps because a key relative is abroad and cannot be reached.

PRESS CONFERENCES

Press conferences should be scheduled regularly, taking into consideration both local and foreign press deadlines. These conferences should also take into consideration the need of elected officials to "be seen."

AMATEURS

In the era of video cameras it has become a known phenomenon that amateurs grab their cameras to shoot footage of disasters. Probably one of the earliest examples occurred seconds after a 1980s bombing in Frankfurt Airport. This footage – piquant and dramatic – can make a strong impression on viewers, but it does not compete with professional reporting. Amateur films do not provide analysis or insight into the incident. If anything, the footage often requires interpretation by a reporter, so that it is placed in proper perspective. In the Frankfurt Airport example, the material filmed was only one very small part of the disaster aftermath; it provided coverage of only one area of the airport terminal.

THE HISTORICAL LEGACY

Some disasters are quickly forgotten. In other cases there is an historical legacy of the disaster and its response. Sometimes the legacy is reasonably accurate, but sometimes the historical memory of a response bears only symbolic connection to real events at the time. Mistakes are ignored and buried. Heroism and efficiency are not recorded, or insignificant players are catapulted to center-stage.

In determining a legacy there are conflicting interests. In an aviation disaster, for example, it is in the airline's interests that the event be forgotten as soon as possible. They realize that a certain amount of publicity and news coverage is inevitable, but all publicity for them is negative. On the other hand, newspapers want to keep the issue alive as long as there is reader interest, and very often the coverage itself generates readers' interest. Coverage, however, does not necessarily mean concentration on real issues; often the articles are human interest stories that have little if anything to due with causes of the incident and an objective evaluation of the response. Again, readers want to be entertained. They also play the game of the fine definition between fascination with the villain's error that caused the disaster and the self-assuring belief that despite the error, it will not happen to the reader.

Sometimes responders want to publicize their own activities, whether they be individual or institutional – and this can lead to conflicting reportage of roles –

if not in deed, then certainly in relative importance. It is from this motive that many articles are written in professional journals (both peer reviewed and not).

Sometimes different legacies exist amongst different audiences. City managers, for example, might remember a disaster response in one fashion, while hospital workers might remember it very differently. Nor are memories necessarily accurate. Many of these memories are influenced by *counterfactual thinking* – the subjective phenomenon of projecting how problems should have been handled (then believing in a revision of reality in which events were handled by the ideal model).

Newspaper coverage does create immediate impressions, but it rarely determines the long-term historical record of an event – until years later a historian analyzes and summarizes source material about an incident. In professional circles it is usually the conference papers, journal articles, and video summaries that fix ideas. Conference papers are often self-serving, highlighting successes and carefully avoiding contentious issues that might become the subject of future courtroom litigation. By the time litigation is over, often years after the event, the legacy has already been molded. Nor are public inquiries an exercise in objectivity – they are usually single-issue focused and politically oriented. Their quasi-judicial nature can be intimidating to witnesses, who also do not usually have the prerogative of speaking off the record. If the final report is objective, it is usually voluminous, tedious reading, and issued far too long after an event to have an effect on the public conscience (Taft and Reynolds, 1994).

Ironically, since the reader has no way of checking the facts presented in the media, he is left to evaluate the ideas and methods presented (whether they were used or not!). The result is the unwitting acceptance of distorted reportage. More often than not, the impression left is an uncoordinated and unverified conglomeration of thoughts and images culled from various newspaper, radio, and television reports.

To cite an example, in one incident a police officer-in-charge wrote in a prestigious publication that one of his first concerns was crime scene preservation following an aviation accident. The theory is perfect, but the reality was quite different. For more than three days that same department actively contended that there was no crime scene; in his opinion the accident was due to a technical non-negligent cause and not due to any criminal act.

In another case a mortuary worker described a forensic medical response based upon the "station" method; perhaps that worker had wanted to use the method, but reality behind the closed doors of the morgue had been very different. There were no "stations"; all responders handled the deceased on the same table, in the same room, in the same place. The best explanation for this is counterfactual thinking.

For many reasons much of the professional reporting of incidents is by pathologists and odontologists. When their articles appear in specialized journals, the influence of these articles is limited. When the articles appear in more general publications (e.g., multi-discipline forensic journals), their influence is much wider. Part of the reason for the difference in influence is that odontology in particular tends not to be indexed in general indices – if indexed at all. Those journals are also more likely to be found in specialized collections rather than in more general libraries, such as those attached to law school. Simply, influence is a function of availability.

One of the reasons that there are not more police articles (hence a better understanding and appreciation of the police role in disaster) is general police reluctance to publish for "security" reasons supposedly tied to "protection of information and methods." It is false to assume that police are any less literate than doctors and dentists; the large number of internal police reports is ample testimony to this ability to express oneself in writing.

LOGISTICS

Handling the media requires allotting them working space, food facilities, and lodgings. It also means consideration must be given to vehicle parking, both of cars and of TV news trucks and tractor-trailers. It also means the press must be provided with the communications needed to transmit their stories.

CONCLUSION

Dealing with the press does not constitute a drain on response forces. The press is an important link with the population. Providing appropriate information should be considered a necessary positive contribution to the response.

SECTION IV

MANAGEMENT ISSUES

Chapter 24
BUSINESS RECOVERY

CHAPTER 24

BUSINESS RECOVERY

INTRODUCTION

Business recovery is the concept of continuing business operations after a disaster. Usually this is thought of in terms of natural disasters, such as hurricane, earthquake or tornado. These disasters can destroy business sites and down communication lines. Businesses, therefore, need programs to continue their operations.

Similar needs exist in transportation disasters. Sometimes facilities are destroyed. In other cases facilities are grossly overwhelmed at the same time as existing staff are assigned to numerous emergency functions. A method is needed to either maintain or restore service to the public.

There are four basic groups affected by transportation-related disasters:

1. Government at its various levels
2. Transportation companies
3. Users/recipients of service
4. Impacted private businesses

Each of these groups has a business recovery concern.

GOVERNMENT

It is a popular misunderstanding that business recovery can be separated from disaster response. According to the mistaken attitude that these are separate concepts, response is considered a government function, and business recovery is considered the concern of the private sector. This simplistic thinking ignores the basic reality that the two concepts are intertwined and interdependent.

From one point of view, government is no less a business than a commercial company. To cite perhaps two of the most obvious examples:

- Revenue collection, whether it be through taxes, licensing fees, fines or other means, needs all of the information back-up systems typical of a commercial business recovery back-up system.
- Government expenditures, whether they be for the acquisition of goods or to pay employees' salaries, are subject to the same basic needs.

It is rare that these factors are directly affected in a transportation disaster, however government is not immune from the effects of a disaster. Government interests *can* be affected. It is clear that government interests are just as vulnerable as those of a private enterprise when disaster strikes. To enable effective disaster response, there is a critical need for government recovery.

In the transportation area, very often airports are government or quasi-government facilities. A simple example of recovery would be to *repair* a runway damaged by a crash. Another example would be to *restore* an airfield fire department to full operational readiness. A more complex example would be one such as the 1998 Swissair crash. For several months the response overwhelmed Canadian government capabilities, from the Nova Scotia medical examiner's office to Royal Canadian Mounted Police laboratories throughout Canada. At the height of the operation the RCMP alone dedicated some 450 policemen to the crash response. Four laboratories (Halifax, Ottawa, Vancouver, Regina) handled DNA examinations (Kerr, 2001). A *business recovery* question is how to give adequate response to the disaster, while at the same time providing *ongoing service* (business continuity) to the general public.

An on-field air crash very much involves both direct and indirect government interests. A direct interest would be operations at the airfield. An indirect interest would be possible tourism decline as a result of an air crash (particularly as a result of terrorism). (Decline in tourism can be from acts of terror in a totally different jurisdiction. Over the years tourism in Jordan has declined when there has been terrorism in Israel, since many tourists canceled their entire plans for what would have been a two-country visit.)

The point is quite obvious, but it is often forgotten or overlooked. The police themselves must have a business recovery plan. The police must continue to function, even if Headquarters must be evacuated as has happened after transportation Hazmat incidents. The emergency switchboard ("911" in the United States and Canada) must continue to answer telephone calls, even if the physical location of the operations room has been shut down and moved.

TRANSPORTATION COMPANY

The main business victim of a transportation disaster is the operating company. Air Florida did not survive the crash of its plane in Washington's Potomac River.

Pan Am staggered along after Lockerbie, until the company finally collapsed. These are just two examples proving the point rather blatantly that an effective recovery plan must be in place, since the very existence of a company may be at stake. That plan must consider such seemingly unrelated issues as company image, continued business operations, and interface with a large number of affected people and agencies.

British Airways is an example of a transportation company that has an in-place plan to deal with overload following a disaster. To free telephone lines for the continuation of regular service, preparations have been made to divert as many disaster-related calls as possible to a special telephone number. Another airline has a program to increase manpower during a disaster by bringing in retired employees (who have the advantage of knowing the company – although there are certainly likely to be changes that have taken place since the date of their retirement.)

SERVICE USERS

There are many businesses that are dependent upon transportation, either for the movement of people (employees or potential customers) or goods (papers or merchandise). These must all be backed up by a business recovery program (Levinson, 1993a).

If a key employee is unfortunately killed in a disaster, the company involved must have a program in place to replace him (despite all of the sorrow involved). A possible client cannot be replaced, but if he is employed by another company, that second company must have a plan in place to have another person take over the function.

The loss of manufactured goods in a disaster (e.g. the Thai International plane in Nepal carried a large commercial shipment of jeans) can most often be covered through insurance. There are, of course, some items that are hard to replace or for which replacement involves a considerable amount of time. These situations can cause business problems for both the supplier and the intended buyer. Much harder to replace can be papers carried. As a preventive measure it is prudent not to carry the sole copy of important papers. A key in business planning is to make copies. A plane need not crash nor a car overturn and burn; the loss of papers can also be non-disaster related, ranging from theft of a briefcase to misrouting of a suitcase.

One of the most common items lost or destroyed particularly in an air disaster is mail. Here it makes no difference whether a company sent business papers by regular post or registered mail. Both methods are vulnerable to disaster.

SECONDARY VICTIMS: PRIVATE COMPANIES

Sometimes a private company that is not a service recipient is affected by a transportation disaster. This can be a business hit by a piece of a falling aircraft or exploding bus, or a commercial establishment closed due to evacuation. In Israel a number of businesses suffered significant damage from debris in bus bombings, even though those businesses were not the intended target (except in the vague sense that "everything" is the terrorist's target). Numerous businesses in the Downtown Manchester (UK) Arndale Shopping Centre never reopened after the 16 June 1996 explosion of a vehicle parked in the area. According to *Shopping Centers Today* (1 November 1998), "Of Arndale's 230 tenants at the time of the blast, only two-thirds remain, the rest having been driven out of business."

The collapse of a floor at the Versailles Hall in Jerusalem on 24 May 2001 is another example of what could also happen in a transportation setting. As bride, groom and invited guests danced at a wedding, the floor gave way leaving more than 300 needing medical treatment (more than 100 of whom were admitted to hospital) and 23 dead. The bride worked in a local bank and had invited many of her coworkers to the wedding. The next morning the bank could not open: 16 workers were injured and three had been killed.

The severity of business disruption should not be underestimated. One company selling business recovery services provides these statistics: only 43 per cent of those companies suffering extensive disaster damage will ever reopen. Of that 43 per cent, only 29 per cent will be in business two years later.

SUPPORTING COMPANIES

Transportation companies can be in unknown danger from a lack of proper response to an incident when usual supporting services are unavailable due to totally unrelated disasters. For example, any transportation disaster, in fact a disaster of any type, is heavily dependent upon adequate communications support.

Following the Loma Prieta earthquake, about 9 per cent of the local cellular telephone network was damaged and out of service – at a time when telephone usage was at a peak. (Just as regular telephone lines can become overloaded during a disaster, the same is just as true for cellular systems.) During the same earthquake, numerous radio and television stations went off the air due to area-wide electrical problems. The same electrical problems caused the *San Francisco Chronicle* newspaper to be printed in limited editions 225 kilometers away. The net effect was that another aspect of disaster response communications was crippled. Although these communications systems are, essentially, private enterprises, they again show the interdependency of the government and the private business sectors in disaster response. If there had been a transportation-related disaster at that time, the lack of adequate communications could well have crippled the response.

RECOVERY PLANS

There are numerous stages in the development of a business recovery plan. After a planning team is established, the first step is to identify possible service problems during a disaster (vulnerabilities). Examples of vulnerabilities are:

- *Equipment.* Some equipment is readily replaceable. For example, if parts of a bomb destroy a restaurant, replacement of glasses and dishes is a relatively simple matter of re-purchasing. Specially made manufacturing equipment can be much more difficult to replace.
- *Inventory.* Again, shelf items can be supplies such as foodstuffs (readily replaceable), or in-house manufactured goods.
- *Records.* This is the most vulnerable part of any business. Many records simply cannot be replaced or reconstructed. A very simple example is the night-time break-in and robbery of a grocery store near the house of one of the authors. The only item stolen was a box of accounts receivable cards. (Each time a regular customer would make a purchase, it would be recorded on a card. Customers would pay once every week or two.) Those records were lost forever. A proper recovery plan calls for duplicate records, with one set stored off-site. The specific type of off-site storage (hot-site, cold-site, shell, etc.) depends upon the nature of the business and the type/use of records.

Is damage to a business far-fetched in a transportation disaster? No. It is not only in the case of urban terrorism that businesses have been affected. Hazmat leaks can affect exposed food in restaurants. Falling debris from an aircraft can hit a business just as easily as a home. Trains have derailed into major power lines that have had to be shut down. Nor is damage necessarily a direct effect of the disaster. Sometimes the disaster response, such as water from fire-extinguishing efforts, understandably causes damage.

Computers deserve special attention due to their increasingly important role in modern business. Computers are also very sensitive and can be affected by heat (from fires) or dust (from an explosion). Duplicate copies of computer products (e.g. narrative material, billing) should always be made.[1] Copies on diskettes are convenient, but again, diskettes must be stored at reasonable temperature, away from the computer with the data. If not, the same disaster can destroy both original and copy.

[1] If for no reason other than computer viruses.

The biggest loss to a business need not be physical damage. It can simply be *down-time*, when the business cannot function. If a restaurant is forced to close because a stricken area has been cordoned off, that constitutes an irretrievable loss of business. If a service-oriented office (such as an employment agency not specializing in walk-in trade) must close, business is not necessarily lost . . . if the company has a recovery plan with preparations to rapidly resume operations temporarily at an off-site location. Moving temporarily versus closing down for a specified amount of time then becomes a decision of choice based upon financial return, rather than a forced closure with no real options.

As in any plan, disaster recovery planning must have the full commitment of senior management. Procedures should be documented, then exercised. The plan must also be updated periodically. If these three factors are not present, even the best disaster recovery plan simply will not work.

Despite the importance of business recovery planning, numerous companies ignore the problem or give it only lip-service. Drabek (2001) cites research that in 1989 fewer than half of the Fortune 1000 companies had developed a company infrastructure to deal with crisis management and business recovery. Ten years later that statistic of fewer than 50 per cent still held true ("in planning stages" considered as still not an operative program).

CONCLUSION

Today business recovery and continuity are real issues when responding to disasters. These are paramount concerns in the commercial world.

SECTION V

RESPONSE TO SPECIFIC TRANSPORTATION DISASTERS

CHAPTER 25

AVIATION

INTRODUCTION

There are many causes of aviation disasters. *Intentional* disasters are generally related to terrorism or some other criminal act. *Unintentional* aviation accidents can be divided into human error (pilot or control tower), health-related incident with the pilot, technical malfunction, or limiting conditions (such as weather). Many accidents are caused by a combination of factors – in other words, a breakdown of quality assurance programs (or a lack thereof).

Butcher and Hatcher (1988) describe a psychological analysis of aviation disasters, calling them not only accidents, but also crises. Use of this terminology is particularly convenient for the authors' psychological orientation, legitimizing examination of psychological reactions to disaster reaction scenarios. One might question if, in the search for culpability,[1] the authors were not placing potential mistake makers (staff in the aircraft, control tower, maintenance crew) on a "public couch" to sort out their feelings.

[1]Very often the search for blame overshadows correction of error to prevent future accidents. Particularly in the aviation field corrections of hardware are relatively common. It is the human systems that are all too often left uncorrected.

ACCIDENTS

The majority of aviation accidents involve private aircraft (including helicopters) or smaller commercial planes. Even when there are fatalities, the number of dead tends to be very few. For example, in 1998 there were 31 accidents involving aircraft of Canadian registry. These accidents took the toll of 87 deaths (TSB, 1999). In terms of disaster response, the accidents (depending upon their location) were very much in the magnitude of traffic accidents. Public attention, though, tends to focus on the minority of accidents – those involving large commercial aircraft (usually on scheduled runs). One accident involving such an aircraft can easily reach disaster proportions and quickly overshadow a long list of small-craft accidents (even with a sum total of more injuries and fatalities).

The response to commercial aviation accidents can create interesting problems, since modern commercial flights often carry passengers from numerous countries. This can mean dealing with an extensive international communications network, taking into consideration time differences, various cultures, and foreign languages.

Air accidents tend to be dramatic events. When two or three hundred people die, this becomes a media event. These accidents have the elements of surprise, suddenness, death, and destruction – all fascinating, but not endangering the unaffected public. The result is that the media consumer focuses on the cost of the disaster (in money and lives), rather than its improbability of occurrence. After all, even by raising *im*probability, the public is shown as potentially vulnerable. (Even in deaths by such common causes as heart attack and cancer, the media consumer is shielded from the medical details of vulnerability.) "Sudden" also shows that there is no pre-warning, as one has in hurricanes; this means that potential responders must command a high level of preparedness, whilst the chances of a malfunction are increasingly minimal.

Flight safety should be measured not by flight miles, but by flight time. Within that framework, current statistics (in contrast to the early days of aviation) show that the most dangerous periods are during the time of takeoff and landing procedures.

PASSENGERS

Sometimes passenger reaction overtakes accident mitigation considerations. Certain safety measures have not been put into effect, because of passenger objection. By the early 1950s it was realized that airplane seats facing the rear of the plane were safer than those facing forward, yet 50 years later we all still sit looking forward towards the closed flight cabin door.

In preparing emergency plans not only should stress be placed on incoming and outgoing flights, but response to overflight aircraft must also be taken into consideration.

Both passenger behavior and public reaction are influenced by preconditioning – the general awareness (usually through secondary and tertiary sources) of air disasters, since very few people are actually involved in air accidents. Those sources tend to be the news media and an assortment of disaster novels and films of differing technical accuracy. Public reaction often compares current disaster incidents and their response to this dubious "experience." These sources become the frame of reference. These sources become the yardsticks of measurement. It is often not the technical report that establishes what happened on the *Titanic*, but rather the Hollywood-produced movie.

HISTORICAL INCIDENTS

The Zeppelins present an interesting footnote in disaster preparedness and response. Of the 145 Zeppelins produced, 143 experienced serious safety problems. For example, the US dirigible, Akron, crashed off New Jersey on 4 April 1933, leaving 73 dead. They were named after Count Ferdinand von Zeppelin (b. 1838), curiously, a volunteer on the Union side of the US Civil War. In 1900 his first balloon-like airship remained aloft 20 minutes, then was wrecked upon landing.

Perhaps the most infamous incident involving a Zeppelin dirigible was the explosion of the *Hindenberg*, designed by Dr Hugo Eckner, that crashed just prior to landing at Lakehurst (formerly Camp Kenderick Reservation), New Jersey on 6 May 1937, leaving 36 dead. Although the incident took place in the early part of the 20th century, lessons can still be learnt today.

One line handler (helping to lower the aircraft) was killed; of the 59 crew members and 38 passengers, 22 crew members and 13 passengers lost their lives. By that date the United States, Great Britain and France had all stopped construction of rigid dirigibles because of their poor safety record; only Germany continued their construction. The crash of the *Hindenberg* (with its two decks, 25 staterooms, and library) marked the end of the dirigible as a means of commercial transportation.

Curiously, the *Hindenberg* had a unique feature unknown in other Zeppelin dirigibles – a smoking room – despite the fact that the 803-foot airship contained seven million cubic feet of highly flammable hydrogen stored in 16 bags. For pre-Second World War political reasons the United States refused to sell Germany helium, which is much safer. To save face the Germans contended that the $600,000 cost of the helium was excessive and not warranted.

Disaster response following the *Hindenberg* was problematic. The airship was scheduled to arrive at 0600, but it was first sighted off Lakehurst at about 1600, when Captain Max Pruss decided to delay landing for at least two hours due to inclement weather. As tired reception staff waited, the *Hindenberg* flew southward to Atlantic City, then returned. After a thunderstorm turned into a drizzle, ropes were tossed at 1922, helping to lower the dirigible to 75 feet off-ground, when there was a fire flicker, and the airship exploded. The net effect was that many of those present had been up since early morning hours, and at the time of the explosion the storm had not yet passed entirely. The explosion was also in full view of meeters-and-greeters, who reportedly reacted hysterically.

New Jersey State Trooper Francis H. Boyer tried to report the accident using a public telephone, only to find that calls by news reporters had tied up the telephone system. Boyer then traveled by motorcycle to Toms River, New Jersey, from where he filed a report via police Teletype. By the time he returned to Lakehurst, the accident site had suffered near-gridlock from arriving emergency vehicles.

Despite the "communications revolution," modern cellular telephones suffer the same problems of overload as experienced at the *Hindenberg* crash site. The same phenomenon of overload took place within 25 minutes after the 4 January 1987 Amtrak crash in Maryland.

Lessons are hard to learn, even when talking about something as straightforward as site access. The Avianca crash in Cove Neck, Long Island (USA) cut off the eastern end of Tennis Court Road (Maher, 1990). The only remaining access was a single long and narrow road with steep drop-offs, which was quickly blocked with the parked cars of responders. Those who arrived later had to make the last two to three miles to the scene by foot. This was further complicated by spectator convergence.

Lockerbie was not the world's worst general disaster, nor was it the world's worst aviation disaster. Why, then, did it have such a strong impact? The reaction to Lockerbie almost makes one ask if activism (lobbying Congress and seeking litigation) is not part of the grieving process. This activism by bereaved families might be considered a more sophisticated variant of street demonstrations and placard carrying. It also allows for the venting of emotion over a more extended period of time than street demonstrations. No one doubts that the political activism has an aspect of psychological release and ventilation. It is also an expression of the wish that the victims should not have died in vain.

RECENT DISASTERS

The United Airlines flight that made an emergency landing in Sioux City, Iowa on 19 July 1989 is an example of a crash *en route*, this time in an unscheduled airport. On 8 January 1989, as the British were still in shock over the Lockerbie disaster, a British Midland flight from London to Belfast, Northern Ireland crashed on the M1 motorway, as it was trying to make an emergency landing at East Midlands Airport.

The crash of Lauda Airlines in a Thai jungle on 26 May 1991 and Thai International Flight 311 on a mountain upon approach to Tribhuvan International Airport (Kathmandu, Nepal) on 31 July 1992 are cases of crashes over non-populated land areas.

The crash of a TWA flight off Long Island (eight miles south of East Moriches) on 17 July 1996; of Egyptair of the US east coast on 31 October 1999; of Air India off the Irish coast on 23 June 1985; and of Swissair off Nova Scotia on 2 September 1998 are examples of accidents over open sea; in all three cases the response efforts were centered in the nearest land areas. Although much of the response was "imported" from other areas, local authorities still carried considerable responsibility, particularly in terms of logistical support and venue security.

Contrary to popular conviction, most air disasters are not weather-related. Departures are delayed when there is bad weather, and arrivals are often diverted. Using US statistics for 1996, only 21 per cent of accidents (30 per cent of fatal accidents) were weather-related. The vast majority of accidents also occurred during daylight hours. Hence, initial response plans are generally activated during daylight and not under extreme weather conditions.

Accidents can be divided into three basic types: (1) mid-air crash or explosion at high altitude with no survivors, (2) mid-air problem where the pilot tries to land, and (3) low altitude or ground crash with injured and fatalities. Most recent accidents are of the third type, although the first two are much more spectacular media events.

FATAL MISUNDERSTANDINGS

This book does not deal with the technical causes of aviation accidents. That is usually a professional subject dealt with by aviation engineers. There is, however, one subject that should be raised, albeit in passing. Communications problems are found throughout disaster situations. Communications breakdowns and misunderstandings are sometimes the cause of the disasters.[2] These communications problems are sometimes the basis of difficulties and complications in disaster response.

A contributing factor to the crash of a Saudi Arabian plane at Riyadh Airport (19 August 1980) was reportedly a language misunderstanding between the pilot and the control tower. The conversation was in English, but language skills were not what they should have been and the pilot was reportedly misunderstood by the control tower.

Point 37 of the Nepal Government Report on the crash of TG 311 on 31 July 1992 cited conversation misunderstanding between the aircraft captain and co-pilot. Part of the problem might have been Point 36, "The co-pilot communi-

[2] Garner (1996, p.10) estimates that 89 per cent of aviation accidents are due to pilot error. A good number of these errors are due to communications misunderstandings.

cated his concern in a mitigated manner." The report also listed as a cause, "radio communication difficulties between the TG 311 crew and the Air Traffic Controllers that stemmed from language difficulties . . . " (Nepal, 1993).

INCIDENTS AND ACCIDENTS

In official aviation jargon an *incident* is an occurrence with no injury or death. An *accident* includes bodily hurt or death. One common light injury is skin abrasion and fractured bones caused by sliding down escape shutes. Roughly 7 per cent of aviation accident fatalities are attributed to burns. A very large proportion of deaths are due to smoke inhalation, similar to the situation in most other types of fires (Levinson, 1993).

EMERGENCY PLANNING

The key to minimizing damage is effective control, yet in a commercial situation that control is an elusive dream. Even the small step of fastening seatbelts is a goal difficult to enforce aboard an aircraft (as noted above in cars). Only when the flight hits air pockets does the vast majority of passengers adhere to seatbelt warnings. The safety explanations at the beginning of a flight are another government rule. The airlines comply by delivering the message, but it is rather questionable how many passengers listen attentively. The content of those preflight safety instructions is also questionable, particularly in light of an NTSB finding (US NTSB, 1985) that most passenger deaths can be attributed to lack of familiarity with safety procedures, leading to an inability to escape the plane (quoted by Garner, 1996).

Normal training concepts would dictate that if there is a possibility that passengers wear life vests, there should be an exercise in putting them on and operating them. This, however, is clearly out of the question given commercial considerations and the desire not to deter customers from flying. The net result is that in time of emergency the airline is not in complete control. The seat belts that are required by law can be left unsecured, and passengers can be at a distinct disadvantage by not being able to use their life vests.

ROLE OF THE CARRIER

It is very basic to say that every airline must have an effective emergency plan, and in fact most airlines have engaged in emergency planning to some degree. That planning is not shared with others. Surprisingly, there is little inter-airline cooperation about what goes into a plan or lessons learnt from a disaster. On the one hand airlines cooperate with code sharing and frequent flyer programs.

On the other hand that cooperation has not been strongly felt in disaster planning, even though it is universally agreed that an accident involving one company can have ramifications that may be felt through the entire industry.[3] A basic reason is the desire to have the incident forgotten as quickly as possible, rather than being re-enacted time after time in seemingly endless meetings.

[3] Star Alliance has begun making some strides in the direction of cooperative emergency planning. The Arab Air Carrier Organization has offered symbolic training, however that is oriented towards course attendees; it does not extend to inter-airline cooperation.

Once an airline has an emergency plan, there must be a specific employee charged with contingency planning and emergency management. He has the responsibility of keeping the plan up-to-date, teaching the plan to employees (including refresher instruction as well as teaching new employees or those with a significant job reassignment), and exercising the plan. This is not a one-time project; it is an ongoing process. This is not a side job for an underemployed clerk; it is an assignment for a professional who understands emergency preparedness and management. Nor is lack of resources a valid excuse not to deal properly with emergency preparations. The tragic results of an airline crash show vividly that proper disaster response can spell the very future of an airline. Disaster response must be a planned activity within overall response, and not an impromptu program designed by the airline after disaster has struck.

Airline emergency plans must be coordinated with fire, police, airport, medical and other authorities involved in possible responses. In the modern world no one can operate as a soloist. What is appropriate in one destination airport might not be feasible in another. This is particularly true when dealing with international destinations.

The overall goal of a carrier is service resumption. Diverting crash response efforts as soon as possible to a separate communications and handling channel does this. It allows regular staff to handle passenger and cargo activities. In other words, the airline must not only have a plan for emergency response; it must also have a plan for business continuity.

From the airline's perspective, one of the reasons an emergency telephone number is circulated is this effort to free the ordinary airline number for routine business inquiries. This is much more pressing at the headquarters or hub airfield where an airline has numerous flights scheduled, as opposed to an out-station where the next scheduled flight after a crash may well be more than a day in the offing.

From a public perspective, a special telephone number gives people a clear address where they can ask questions about family and friends who might be involved.

There is no one answer to establishing an airline information center, but several basic rules can be followed, as outlined below.

Toll-free or local call telephone access should be arranged for use in the origin, destination, and site-of-disaster (if different) countries. An Internet site was opened after the October 2000 crash of Singapore Airlines in Taipei,

Taiwan. This has the disadvantage of being cold and impersonal. While it is efficient to convey formal press releases, it should be used very sparingly to convey personal information such as passenger lists.

There should be full foreign language service provided at the information center, according to the nationalities aboard the flight.

It is best for telephone calls to be first answered by a recording stating the date and place of the crash, the flight routing, and the flight number. This short step has the advantage of reducing the number of calls to be answered by providing basic information to people uncertain as to which flight was involved.

Once it is established that a caller is inquiring about a specific person aboard the flight in question, one of the most important tasks of personnel answering the telephone should be to receive information. For this purpose they should be given forms with a short series of questions, including the relationship of the caller to the passenger and contact instructions. The form should also contain appropriate office routing options (missing persons, medical information, property, etc.). After it is established that a caller is next-of-kin or a close relative to a person aboard a crashed flight, that caller should be given a new telephone number for future telephone conversations so that he can bypass the initial filtering process.

Access to the information center should be carefully restricted to duly authorized personnel, so that information can be protected and a working atmosphere maintained. Access to bereaved families should also be restricted. Following the Egyptair crash a newspaper reporter was arrested after he feigned to be a bereaved relative in order to interview next-of-kin.

British Airways has taken a major step in emergency planning by establishing its Emergency Procedures Incident Centre (EPIC) at its Heathrow Airport facility. EPIC is designed to be the information communications hub of the airline (with services contracted to other carriers) following an air disaster. The center is equipped with lines for both incoming and outgoing communications, with pre-assigned working stations for those groups (e.g. police, medical) in liaison with air carriers in times of emergency. The intention is that EPIC be activated any time an aviation accident is related to London (departure, arrival, large number of passengers, etc.).

Despite all that has been written, there are numerous cases in which airline employees did not do what was expected of them at the time of a crash. In one current case an airline office at the airport of flight origin was promptly closed when it was learned that the aircraft had crashed upon landing. Families in that originating city had no direct local source of information.

AIRLINE CHECK-IN PROCEDURES

The growth in air traffic over the past 20 years has meant that airlines have had to keep pace with technology and create new methods to handle passengers and assure their security. These new systems can be of direct assistance to disaster responders when there is an accident.[4]

Today passengers are no longer anonymous bearers of handwritten tickets. The tickets, issued as part of a *common user system* (CUS), are pre-recorded as part of an airline's *departure control system* (DCS). It is through these computerized records that the travel agent issuing the tickets can be quickly located, as well as record of payment and very often personal details such as address and telephone. Checked baggage is tied to the passenger through a *baggage source message* (BSM) and its tags, affixed to both ticket and luggage. The baggage is loaded into a "bin" or unit load device (ULD) which is loaded aboard the aircraft. After passengers board the aircraft, there is a reconciliation of passengers (by boarding pass) and luggage to determine flight load.

The security implications of this system are wide-ranging (Pearson-Fry, 2000), as is the meaning for disaster responders. Computer tracking systems, when used fully and properly, provide substantial proof that a given person was aboard a flight. Finding baggage in wreckage reinforces that proof. The system also gives substantive assistance in linking tickets on a multi-flight trip and locating possible next-of-kin.

Passports filled out with optical character recognition (OCR) fonts pose problems to supposed "spouses" when boarding international flights. These passports are often scanned as part of the check-in process. Such "couples" (usually a married man and not exactly his wife), have been discovered on passenger lists in more than one disaster.

Many people have citizenship and passports from more than one country. This should be taken into consideration in disaster response.

[4] Despite the availability of security measures, sometimes even the simplest systems are ignored. Per the Chief of Police at Heathrow Airport, there was a reconciliation mistake on the Pan Am 103 flight that crashed in Lockerbie. One person whose baggage was aboard the flight arrived at the boarding gate after it was closed. He was soon cleared of any criminal suspicion.

JURISDICTION

Jurisdictional questions arise in many incidents. On 24 November 1993 a Bell-206 helicopter crashed into an open field north of Beit Qama, Israel (Israel Civil Aviation File No. 10/93). The first police units to respond were from the Negev Sub-District. Soon, however, it was realized that the helicopter crash was less than one kilometer outside their jurisdiction. At that point representatives were summoned from the Lachish Sub-District and told to assume command.

Another example of jurisdictional problems posed by an aviation disaster occurred on 13 January 1982 when Air Florida flight 90 crashed into the Potomac River in Washington, DC, shortly after take-off from National Airport. Weather conditions were poor with low visibility, high winds, and snow falling on the ground. The first authority to assume jurisdiction was Arlington, Virginia. This was followed by Washington, DC (at first the fire department). Unlike most rivers that divide jurisdiction in the middle of the water, the entire Potomac belongs to Washington, based upon the nineteenth-century cession of what are now Arlington and Alexandria to Virginia. The Potomac and the crashed aircraft were better accessed from the southern/western shore of the river (i.e. Virginia). A situation thus arose in which the District of Columbia exercised its jurisdiction from bases in Virginia, outside its territory with mutual aid (fire, medical and boats) from neighboring areas.

The Air Florida plane had been cleared for takeoff at 1559, and it fell from an altitude of 350 feet after a 30-second flight. At 1604 the DC Harbor Police received an unconfirmed report of an air crash. By 1620 a US Park Police helicopter[5] was over the downed aircraft and rescuing four of the five passengers who would survive. (Seventy-eight people aboard the aircraft were killed. Four people on the ground also died.) Only at 1622 were air and ground responders working on the same communications frequency (DeAtley, 1982). Space at the Air Florida site allowed for the operation of only one helicopter.[6]

[5] Eagle 1. Other helicopters from Ft. Belvoir were ready for deployment, however the crash site did not allow for more than one helicopter to maneuver (Emergency Department News, February 1982).

[6] A similar situation existed in the response to the AMIA Building in Buenos Aires. The narrow street next to the building was the basic factor limiting the response, and not the manpower or equipment available.

Air Florida is an example of a multiple-type disaster in which one incident served as a catalyst for another. Cars caught in the snowstorm traffic jam on the Fourteenth Street Bridge were hit by the Air Florida plane. As the focus of activity was on the aircraft, the Metro subway system was also experiencing difficulties. At 1629 Metro Train 410 struck a concrete wall and derailed as it was being removed from a rail approach to the Air Florida crash area. Underground communications limitations and problems in shutting off power in the third rail further complicated response. By 1709 response personnel and medical teams were being transferred from Air Florida to handle the accident in the Metro system. Only at 1830 was the last injured passenger evacuated to hospital.

Hospital coordination at Air Florida was problematic, posing serious questions. The Washington Hospital Center, four miles from the crash, did not receive notification despite a standing procedure. When a senior hospital official heard radio reports, he put the hospital on alert. In fact, only one patient arrived; she was immediately moved to an operating room to make room for the other injured (who never arrived).

Life-saving operations were halted at 1800 hours in light of a professional determination that there was no further possibility of saving lives. For aircraft passengers that was true, but at 2300 another victim was spotted on the bridge. The woman, suffering from shock, had abandoned her car on the bridge and stood in the blistering cold. A proper search of the area had obviously not been done.

On 13 January 1982 National Airport Manager, Augustus Melton, testified before the National Transportation Safety Board, and described the airport disaster response as satisfactory. Factually, he was quite accurate, but in proper perspective the airport had only a minor role to play, since the accident took place off field and outside its legal jurisdiction.

AIRCRAFT AND BODY RECOVERY

Usually at a crash site the accepted method of plotting the scene, where possible, is to place ribbons to form a grid, labeled horizontally by number and vertically by letter (Haertig, A., 1990), similar to work that is done at an archeological site (from where the technique was borrowed). At numerous sites, such as the crash of Air New Zealand at Mount Erebus, Antarctica, different color flags were placed to denote airplane wreckage, property and human remains (Taylor and Frazer, 1980). This method is obviously difficult to apply when a plane crashes into water.

It is not infrequent that airplanes crash in places difficult to reach. On 4 September 1971 the remains of Alaska Airlines 1866 were found after an extensive search 18.5 miles from Juneau (Alaska, USA) Airport on the side of a 2400-ft ridge. Due to steep slope, hoisting equipment had to be used to recover 111 bodies, property, and aircraft (Chapple, 1972). Working conditions were trying, and it took five days for the scene to be photographed and plotted. A program had to be devised to protect the bodies – from wolves, bears, wolverines and other scavengers. (Despite the difficulty to reach the site, it was still necessary to arrange for a press pool of four to visit the crash area.)

The crash of Thai International on its landing approach to Kathmandu was in an even more difficult mountain terrain. In this case it was necessary to trek four hours from the forward helicopter landing area. This meant that only the most important items (human remains and lightweight property) could be recovered. Heavier parts of the aircraft had to be left *in situ*.

GOVERNMENT INVESTIGATION

GREAT BRITAIN

In Great Britain the Department of Civil Aviation was established in 1918, and under its authority the Accidents Investigation Branch (AIB) was set up. Its first investigation was a non-fatal crash on 31 May 1919. The first fatal crash to be investigated was on 24 December 1924. Over the years the Department of Civil Aviation has been reorganized and renamed under a variety of administrative and legislative actions, however the principle of governmental investigation of accidents has remained.

In England questions regarding the medical condition of a pilot involved in an aviation crash date back to the 1920s, and in that period there are occasional reports of autopsies. It was not until the 1950s that autopsy and toxicology became standard practice in aviation accidents in England (Stevens, 1970, p. 3).

UNITED STATES

According to the Air Commerce Act of 1926 the US Department of Commerce was charged with the investigation of air crashes. This responsibility was transferred to the Bureau of Aviation Safety, then after the establishment in 1967 of the Department of Transportation, the National Transportation Safety Board (NTSB) was set up. In 1974 NTSB was separated from the Department of Transportation and given independence to study the cause of transportation accidents with recommendations for increased safety. Although, as the name suggests, NTSB works industry-wide, its highest profile investigations are usually airline-related (comprising about 50 per cent of NTSB investigations).

NTSB investigators can examine foreign accidents under treaty agreements with the International Civil Aviation Organization (ICAO).

According to the US Aviation Disaster Family Assistance Act of 1996, NTSB was given the responsibility to coordinate federal aid to aviation disaster victims and their families. Airlines continued to retain responsibility for notification[7] to families, but NTSB was charged with keeping families up to date with investigative information, including victim recovery and identification.

Only in 1954 did the autopsying of the cabin crew become standard procedure following a fatal accident.

CANADA

The Canadian Transportation Safety Board (TSB) was established as an independent agency in 1990, with the objective of advancing transportation safety.

[7] There are serious questions about the ability of airlines to notify families. Although passenger lists provide critical information for that notification, airlines simply do not have the personnel (and usually not the training) if passengers are from numerous places. The main exception is charter flights that tend to have people from a more limited area. Even then many carriers are hard-put to find enough personnel to make notifications. Telephone notification is *not* recommended; amongst other problems it creates a situation of no control on the recipients' reactions.

TSB has the authority to conduct investigations; its jurisdiction is in Canada and abroad when dealing with transportation means of Canadian registry – making them another player concerned with the disaster response.

The net result of this situation is that following an aviation accident both the interests of aviation investigatory agencies and disaster response teams must be taken into consideration. At the scene this means police/aviation coordination in looking at technical evidence to determine technical cause and criminal culpability. In the morgue there can be situations where aviation industry authorities require autopsies of crew and passengers to determine the course of events (Cullen, 1980), even though mortuary staff do not need the autopsy results to identify victims.[8] Additional reasons for autopsy might arise from death amongst family members to determine inheritance sequence (Cullen, 1980) and possible suffering before death as a basis for suit by surviving families (Salomone, 1987).

[8] Autopsies on aviation fatalities are recommended, although not mandatory, in ICAO Aircraft Investigation Manual, Annex 13 (11 April 1951 and revisions).

SURVIVORS

Most people do survive air crashes (NTSB, 2001), although they might well suffer injury.

For survivors an aviation accident is very different from many other disasters. In a large number of cases the survivor is far from his usual environment. Family and friends, who can constitute a normal support group after trauma, may be far away. There can also be strong cultural differences between "home" and the accident site, making readjustment all the more difficult. This should be taken into account when arranging for psychological assistance.

The pilot of an aircraft that has crashed is in a very difficult situation. His actions will be scrutinized quite carefully, even in cases where it is quite apparent that the cause was technical. This review of the pilot's performance can be an inhibiting factor in his applying for psychological assistance; pilots are afraid that such requests for help can interfere with their career future (Butcher and Hatcher, 1988).

When travelers are making their trip with family or close friends, it is only normal that survivors inquire about others with them on the trip. Responders should anticipate these questions. Sometimes the answers are straightforward and simple. Sometimes the answers must be given very delicately (as in the case of serious injury or death), and sometimes the questions are asked before answers are available. In an air crash, for example, it simply takes time to reconcile the list of passenger names with the location of each person. It is not unusual for the uninjured to be in a rest and debriefing area, while injured people are in hospital.

Thought must be given to notifying survivors of the fate of others in their group, especially when injuries are severe or fatal. This is in addition to notification to close relatives or next-of-kin.

THE DEAD

It is important to record the seating position in which bodies were found. This is required for accident reconstruction and for later identification (even though the seat number is only an investigative lead and not an absolute proof, since passengers may change seats, even in a crowded flight).

An air crash can result in a large number of bodies exceeding the refrigerated space available in a morgue. When properly cooled containers are used to augment space, all commercial writing should be painted over or otherwise blocked from view on the van, particularly if the container belongs to a food or shipping company. If the name of the van owner is known, there can be significant reluctance to use the van after the incident, despite meticulous cleaning.

Air Inter crashed outside of Barr, a small town near Strasbourg, France. In that case bodies were kept in special tents equipped with insulation and refrigeration. They were stored in these tents, then brought several at a time to the Institute of Forensic Medicine in Strasbourg for examination and identification.

BEREAVED FAMILIES

In many *Western* cultures bereaved families want to visit the disaster response area. From a professional point of view this has the detrimental effect of distancing relatives from their homes – where much needed *ante mortem* information about the victim is often to be found (Doyle and Bolster, 1992).

In certain *Eastern* cultures (e.g. Japan), the custom is to travel to the exact site of the disaster, then return using a circuitous route to rid oneself of the evil spirit. From an information perspective the net result is the same. Again, the family distances itself from much needed *ante mortem* records.

DEALING WITH AIRCRAFT FUEL[9]

[9] This is written with sections courtesy of CONCAWE (Bruxelles) from Report 99/52 and with the assistance of Eric Martin of that organization.

The primary danger from aviation fuel is fire. Although the fuel is not as easily ignited as gasoline, it certainly does burn. As testimony to this fact, there is no shortage of aviation crashes accompanied by fire.

Fuels used in commercial aviation fall broadly into three main types:

- kerosene type, usually blends of different kerosene components;
- "wide cut" type, in which kerosene components are blended with low

flash point naphthas, for example, heavy straight-run naphtha, to give more volatile fuels covering the C_4 to C_{16} carbon number range;
- high flashpoint kerosene type, blends of kerosene components having a minimum closed cup flashpoint of 60°C.

The main grades of jet fuel are as follows:

- Jet A: as Jet A-1, but with freezing point of –40°C. Available only in USA and Canada.
- Jet A-1: kerosene type fuel used in civil aircraft. Maximum freezing point of –47°C.
- Jet B: wide cut used in civilian aircraft.
- JP-4: wide cut used in military aircraft.
- JP-5: high flashpoint kerosene type used in naval aircraft.
- JP-8: kerosene type used in military aircraft.

Jet fuel has a fairly low acute toxicity. Even so, it is recommend that both skin contact and vapor inhalation be minimized, although toxic hazards are by no means the primary hazards to rescuers and survivors after an air crash. Avoiding skin contact and particularly inhalation are, of course, not always possible. The prevalent opinion is that inhalation toxicity does not warrant rescuers wearing special masks.

Fire is an obvious hazard in dealing with *any* type of fuel. Fuel is a major concern of airport fire departments. One of their functions is to supervise the transfer of fuel from arriving trucks to storage facilities to departing aircraft. In most countries there are local regulations regarding the frequency of preventive equipment inspections to lessen the chances of accidents.

It should be remembered that aviation fuel poses a hazard not only in airport storage tanks and aboard aircraft, but also in its various stages of transportation before it even gets to the airport. Fire brigades along transportation routes must be equipped to extinguish fuel fires (again, not restricted to aviation fuel). In practical terms this means the use of not only water to put out fires.[10]

[10] Aviation fuel fires are usually extinguished using Aqueous Film-Forming Foam. AFFF works on a principle of separating fuel from oxygen by various means.

On 7 July 2000 an aviation fuel tank truck on its way to Delhi collided with a truck near Baghana, Rajasthan (India). Three persons were burnt alive.

In the Lufthansa crash in Warsaw the aircraft fuel tanks broke, creating a fuel spill. The JA-1 fuel was probably ignited by overheated parts of the damaged left engine caused by an electrical short. Whatever the reason, the result was conflagration in an area of about 600 square meters; then the fire penetrated the passenger cabin.

Fuel, however, is just one cause of aircraft fire, as the China Airlines example shows. Yet another cause of fire aboard an aircraft is the extensive electrical system found on most commercial airlines. Electrical fires tend to be relatively small and smoldering, until they ignite another material (such as fuel).

On 26 April 1994 China Airlines Flight Number 140 (A300-600R Airbus) from Taipei International Airport crashed upon landing at Nagoya Airport at 2015 Japanese Time. 264 of the 271 people aboard were killed.

According to the accident investigation report, "Fire broke out, and flames as high as a three-storied building enveloped an area more than 100 meters wide." This was caused not only by jet fuel, but also by cargo carried on the flight. The fire was totally extinguished only at 2148, 93 minutes after the crash. The fire had, of course, implications for the Search and Rescue operation, complicating the entry to the plane of emergency response personnel (Ladkin, 1996).

Besides fire, jet fuel poses a relatively small danger to the environment, because (1) it evaporates relatively quickly, and (2) it is almost insoluble in water and floats on the surface of water (hence it causes minimal underwater damage). In most cases jet fuel is left to evaporate or is skimmed off the surface of water. On land free liquid is collected, then efforts are made to encourage evaporation and biodegradation.

MISUNDERSTANDINGS

Sometimes there are misunderstandings, even during a fire. In the 1987 crash of Air West the cabin crew stationed themselves at the aircraft doors to assist passengers in leaving; the fire department thought that these people at the doors were blocking the exits and ordered them to move.

ROLE OF THE AIRPORT

Just like airlines, airports must also have emergency plans (Barbash *et al.*, 1986), coordinated with the plans of those airlines servicing the field and with other relevant responding agencies. The airport plan, for example, must include a wide range of services ranging from increased telecommunications to companies supplying portable toilets.

It is insufficient to write a plan just to fulfill a technical requirement. The plan must be thought out properly. An example of poor and uncoordinated contingency planning is one airport visited by the author; it was planned to use the baggage retrieval area for meeters and greeters, totally ignoring the lack of tables and chairs (or even the room to bring in furniture from elsewhere).

In the United States the requirement for an emergency plan at a certified airport is governed by FAR (Federal Aviation Regulation) Section 139.325. That plan must include response to various types of disaster (aviation, bomb, fire, natural, etc.) as well as different types of response activities (removal of injured and uninjured passengers, aircraft removal, etc.). Although the most common airport disaster involves only aircraft, such scenarios as a crash into a building, a terminal bombing (as happened in Frankfurt) and so on must be considered.

One basic function of the airport is to provide immediate response facilities. This means, as noted above, place to receive the "meeters and greeters" associated with a flight. Their reception area should include facilities to serve light refreshments, telephone lines to put at their disposal, casual seating, and table space for interviews with police and accident investigators. Many airports have contingency contracts for the purpose with nearby hotels.

After a plane crash, attention is generally focused on high-profile response activities at the airport or at the crash site. Response, however, has to take other areas into consideration. One of the most important functions that cannot be overlooked is the rerouting of ground traffic so that emergency vehicles can enter and exit the crash area. This includes local police enabling vehicles to reach pre-designated staging areas and to move forward when instructions are given (Levinson, 1993). This also means enabling fast movement of emergency vehicles on routes to hospitals, and controlling traffic at those hospitals.

When an airport authority fails in disaster response, the problem generally becomes that of the airline. This is true when a plane crashes in flight, distant from an airport. After all, the public does not designate an airport as "unsafe"; people will, however, speak of an "unsafe airline," whether the epithet be justified or not.

EXERCISES

Despite the public perception associating air crash contingency preparations with the carrier, that is not the complete reality. Airlines are responsible for their flight safety and prevention of accidents due to actions of the crew or condition of the aircraft. Airlines, however, are not responsible for the emergency response to crashes. They might run their own internal exercises or emergency training programs, but they do not initiate emergency exercises with responding agencies. At best, they participate in those exercises. Some airlines have the attitude that if the public sees an airline participating in an exercise, this will instigate fears of flying or memories of previous crashes.

For historical reasons aviation exercises are considered to be the responsibility of airports. However, many air crashes are outside the airport perimeter. In most cases the legal authority of the airport authority is limited to the perimeter fence, or a very small and clearly defined area beyond it. This means that the airport has no legal response mandate. Outside the airport perimeter any response is voluntary, usually based upon economic and humanitarian considerations. Sending certain equipment such as fire engines usually means closing the airfield to arriving and departing flights.

On field the airport *must* respond, even if precautionary deployment means closing of the field. The responsibility is spelled out both in local laws and ICAO regulations.

Since exercises have become airport activities, they are, therefore, conducted at the airports (which are required to have all response functions), rather than at distances from the airport, where many accidents actually take place and where immediate response resources are not necessarily available.[11]

Another frequently ignored aspect of disaster response is international cooperation. Airport exercises very rarely include the "other end" of a flight. When there is a crash of a commercial flight in or near the airport of departure or arrival, there is always activity in the other airport involved. That activity involves both the sending and receiving of information between the two points. That exchange should be the subject of exercises.

There are many basic differences between an on-field crash and one outside the perimeter fence in a populated area. One difference is that on field, even if the plane breaks up and there is more than one crash site (e.g. United Airlines at Sioux City), there is little secondary damage (unless the plane crashes into the terminal facility). A crash into a populated area can mean ground casualties, fires, broken gas lines, etc. This means that in addition to dealing with the aircraft – a major undertaking in itself – there must also be a response to the ground situation. Despite this operational complexity, such situations are very rarely the subject of exercises.

It is desirable to carry out exercises on a frequent basis. This is due to turnover in personnel and the usual lack of previous disaster experience. A proper airport exercise is not an isolated event; it is the annual or bi-annual[12] culmination of a long training process (Cheu, 1976). That process starts with writing or updating emergency plans. In that first step, the airport reviews its emergency procedures to see if something is missing, or if there are lessons to

[11] At the time of writing this issue is being addressed on a very constructive basis at Boston's Logan Airport.

[12] Per ICAO and national requirements in various countries. Although the exercises are required, their content is quite "flexible" and not set forth in any real detail. Sanctions against airports not complying with these vague requirements are also virtually unknown.

[13] One main problem with lessons learnt is that all too often they are self-serving for the initiating organization, or they are donor-oriented rather than recipient-oriented.

be learnt[13] or changes in the airport to update existing procedures. Only then can exercises be contemplated.

Usually a period of three months is the bare minimum needed to put together an effective airport exercise. During this time decisions will be made regarding what response functions should be tested. Once that has been determined, it is then possible to decide upon the type of emergency, aircraft, location, number and types of injuries, cargo, etc. The basis of exercise scenarios can be previous incidents, local or otherwise. It is desirable, however, to introduce new elements – potential disasters that have not yet happened – to truly test procedures and response.

There can be different types of exercises:

- *Dry*[14] *exercise.* In this exercise participants are given a series of events in an accident or emergency, and they are requested to suggest reactions. In a well-constructed scenario, various complications will develop as the exercise progresses. This type of exercise can be used to gauge knowledge of procedures, response options, and general disaster thinking. A dry exercise is often a preliminary stage to a live exercise.

  One method used to heighten interest is to let events develop at a real-time pace with a clock setting a deadline for answers. Sometimes it is necessary to skip periods of time to exercise elements of the scenario that happen at a much later hour.

  Alexander (2000) suggests a scenario with a context of hazard, vulnerability, and risk; then there is an event with impact, reaction, and secondary reaction.
- *Partial exercise.* Air disaster exercises can be costly, both in equipment and in manpower. Sometimes exercises highlight just one or two issues. This can be a cost-effective way of solving specific problems.
- *General exercise.* This is sometimes called a "full exercise," however this is a basic misnomer. No exercise can or should try to include all aspects of air disaster response. Simply, the range of activities would be much too broad for meaningful examination and criticism. For this reason a well-designed exercise tests out a pre-defined list of responses[15] within the functional areas described above. One exercise might target fire fighting, rescue, and medical triage.[16] Another exercise might deal with reconstruction of the passenger list. Subjects rarely exercised but very much needed are fatality handling and identification, coordination with "people-oriented operations" in the origin/stopover/destination airports, and telephone inquiry systems.

[14] Also known as "table top."

[15] Common areas to be tested are rescue, fire, medical, airport operations, communications (including mutual aid), security, command/control.

[16] Disaster exercises often include triage, since it is a field procedure not used in non-disasters (even though a version of it is often used in emergency rooms to determine priority of treatment). A practical result of triage's not being a standard procedure can be seen in the evacuation of some victims of the Hyatt Regency skywalk collapse in Kansas City; one helicopter company innocently ignored triage/hospital assignment and transported the injured as the flight staff saw fit. According to Stout and Smith (1981), triage had not been part of any pre-disaster plan.

There are numerous methods of running a live exercise. One approach is to simulate injury with theater-like make-up. Another approach is that doctors do not need to practice identification of injury in the context of an airport drill; signs or labels indicating injury are sufficient. In truth, both approaches are valid. The key is in the planning decision – what aspects are being exercised. One pragmatic consideration is the financial cost of running the exercise.

Many airports produce an edited film of their exercises. The version that is released to the public might be interesting to the uninitiated, but it is essentially a public relations ploy helping to build the image of the airport. Such a film is never a critical analysis of what "went wrong" at the exercise; it is a public commercial of what "went right" in the context of concern for the flying public. A hint of this approach can be found at one US exercise in which the second-in-control was the Public Relations office.

The unedited video footage is important. When an exercise is planned properly, viewing that footage can be a chance for critical evaluation. The problem is that very few responding agencies have access to the unedited film.

An airport exercise is a good opportunity for airlines servicing the field to check co-ordination of their emergency procedures with those of the airfield. Therefore, it is desirable for the airport to consult with airlines as exercises are planned, so that those elements of interest to the carriers can be included. To cite just one example, in September 2000 American Airlines used the Logan Airport (Boston) Airport exercise to test its own emergency procedures.

It is not infrequent that the true lessons to be learnt from an exercise relate to general management principles, such as resource management, management–employee relationships, shift planning, etc. These lessons are often overlooked in a micro-examination of the exercise, where only seemingly disaster-specific details are examined.

A common method to assist in lessons learnt is preparing a checklist which exercise evaluators must complete. These forms should be function-specific with detailed evaluation questions, and not so general that the forms can apply to every and all aspects of the exercise.

Another benefit of exercises is that they tend to build up the self-confidence of participants. If the exercise is successful, participants will have a stronger feeling that in a disaster they *know* what to do and they *can* do it.

Every exercise must have a provision to switch from exercise to real situation. This switch can be for a supposed victim who has a real medical emergency, or for an aviation or other situation (e.g. a major disaster that requires the services of medical personnel assigned to the air exercise) that develops as the exercise is being held.

AIRCRAFT AT SEA

There is no end to possible disaster scenarios. On 22 March 1992 a USAir Fokker F-28 crashed into the perimeter fence at New York's La Guardia Airport. According to reports, helicopters and rescue boats caused choppy seas, complicating the rescue of those passengers who were forced into the waters. Police boats, designed for patrol, were built with sides too high for sea rescue. This crash produced an interesting problem of jurisdiction. According to understandings in effect at the time, a crash inside the fence was under the response jurisdiction of the airport. Outside the fence local police were in charge. In this case the plane was divided by the perimeter fence; theoretically, two authorized jurisdictions were in charge over different parts of the aircraft. It was decided that since the airport had plans specifically for an aviation accident, they would be in charge of the entire operation.

Aviation disasters at sea are a frequent occurrence. Just from 1996 until this time of writing three commercial aircraft have crashed at sea after taking off from J.F. Kennedy Airport in New York: TWA (17 July 1996), Swissair (2 September 1998), Egyptair (31 October 1999). In many respects these accidents are treated as regular aviation disasters, however they do have certain unique aspects.

In such a crash, locating the exact impact site can take time, since the aircraft is often totally submerged. In the case of Swissair this took some ten hours. The situation was complicated by night darkness, rough seas, and a need to organize forces (primarily volunteer).

The sighting of floating debris does not necessarily indicate the exact crash site, since as the descriptor indicates, the debris *floats*. What must be done is to calculate factors such as tides, winds, and time since impact to determine from where the debris floated, with a projected float pattern. (Air disasters usually leave extensive amounts of floating items, and it can take considerable time to collect all pieces. It is necessary to chart both past and future float to properly deploy collection ships.)

Part of the search for Swissair was done by a DC-3 and a Convair-580 aircraft owned by Environment Canada. The two planes were equipped with special equipment to locate debris and fuel spills. Part of the search was also to answer the investigative question as to whether the pilot had dumped jet fuel to lighten his load as he tried to make an emergency landing (Brown, 1999).

Without any doubt the first priority is the search for any survivors. A common immediate danger to survivors in the water is hypothermia. The colder the water, the faster the onset of hypothermia. The more obese a person, the more "protection" he has against hypothermia. It should be remembered that airline passengers do not have access to many lifeboats; in most cases they have life vests, which means that they are in direct contact with the water.

Experience has shown that volunteers searching for survivors (often private yachts or fishing vessels) are psychologically unable to cope with the collection of floating bodies, and particularly body parts.

It is critical to determine the cause of an accident so that future accidents might possibly be prevented. Although part of the information is often quite technical, it is often important to know what transpired in the passenger cabin. If there are no survivors to describe what transpired aboard the flight, sometimes significant details can be learned through pathological examination (assuming there are more or less intact bodies).

Commercial aircraft carry two "black boxes" (an historical term, and today in fact often an easily sighted fluorescent orange and not black) – the Flight Data Recorder (FDR) and the Cockpit Voice Recorder (CVR), usually equipped with a locating signal.

There can also be complications when aircraft crash into shallow water. In Nigeria an internal flight crashed into a swamp and sank "only" twelve meters; extrication and recovery were only partially successful. Mud is often raised in clear water, complicating underwater visibility until it clears.

CONVENTIONS

There are a number of international conventions concerned with aviation accidents.

Chicago, 7 December 1944. Signatories must allow peaceful overflight and emergency landings. As amended, serves as the basis for *ICAO Annex 13* (recommended accident investigation procedures).

Warsaw Convention (12 October 1929). Amended by the *Hague Protocol* (1955). Additions contained in the *Guadalajara Convention* (1961). These documents set legal and financial liability limits.

The Warsaw Convention has been recently supplemented by a series of legal agreements signed by many airlines based upon the *Intercarrier Agreement* of the International Air Transport Association (IATA). These agreements waived the damage limitation of the Warsaw Convention. According to the agreements carriers are now liable for full compensatory damages in cases of passenger death or injury in an accident, unless it can be shown that "all necessary measures" were taken to avert said damage.

Until recently liability was generally set at $75,000. That sum has now been raised. Following the 1998 crash of TG 261 the families of deceased persons were offered $100,000 per person killed. SilkAir offered $140,000 to families of deceased passengers from their flight that crashed in 1997; KING Channel 5 Seattle reported on 28 April 2000 that actual settlements ranged between $144,000 and $200,000.

In the United States the Congress passed the *Death on the High Seas Act* (DOHSA) in 1920, limiting damage to economic loss only. DOHSA applies to accidents "on the high seas," and not to accidents in territorial waters. The law also applies to airlines as clarified by an April 2000 amendment. Thus, the Egyptair Flight 990 disaster that occurred over international waters is covered (and damages restricted); the TWA Flight 800 and Swissair Flight 111 disasters in territorial waters are not covered by DOHSA.

CONCLUSION

Disaster response is a major concern in the aviation world, since air accidents are high-profile and attract heavy public interest. Even so, many aviation companies are not as well prepared as they could be. Response to accidents at sea needs no less advance preparation, even though they do not attract so much public attention.

CHAPTER 26

DISASTER AT SEA

INTRODUCTION

Disasters at sea have plagued society from ancient times. Today, however, mitigation is more effective given increased knowledge in navigation and meteorology, as well as better means of communications and dispatching of aid. Even so, travel by ship can be hazardous, even over short distances.

HISTORICAL NOTE

Perhaps the world's most popularly cited modern civilian naval disaster was the sinking of the *RMS* (Royal Mail Ship) *Titanic* (White Star Line) on 14–15 April 1912 after colliding with an iceberg in North Atlantic waters. In all, 1503 persons lost their lives. Virtually all those saved were picked up from lifeboats, underlining the importance[1] of these small vessels.

One curious historical note is that three responding ships departed Halifax, the nearest convenient land point, to deal with the disaster area. Two of these ships managed to recover a significant number of bodies – so many that the ships could not return to harbor safely carrying the deceased passengers of the *Titanic*. The bodies were assembled, a memorial service was held, then bodies were returned to the sea.

[1] It has been suggested that the number of lifeboats on a ship should not be an absolute number based upon vessel occupancy. That number should take into consideration other factors such as availability in different parts of the ship and less than maximum usage of each lifeboat.

PASSENGER SHIPS AND FERRIES

Although travel by sea is no longer as common as it was before the advent of aviation, sea disasters still do occur, both with passenger and cargo vessels. There still are numerous cruise ships, but ferries also carry large numbers of people. Over the years there have been numerous ferry disasters.

The fire on the *Noronic* (Brown, 1952) on 17 September 1949 serves as an example of a disaster on board a ship in harbor. The 37-year-old ship caught fire at port in Toronto. Of the 527 passengers aboard (mostly from the Cleveland and Detroit area), 119 died. Of these, 116 were eventually identified during a four-month operation, despite extensive damage (only 16 bodies had all four limbs). The fire had broken out in a linen closet and spread rapidly throughout the ship, blocking the gangway to the pier.

The sinking of the *Herald of Free Enterprise* outside the Belgian port of Zeebrugge on 6 March 1987 raised significant issues (Bruggeman, 1992): as is usually the case, the ferry had no passenger list, hence it was difficult to determine who was on board.

On 7 April 1990 the *Scandinavian Star*, a ferry en route from Oslo, Norway to Fredrikshavn, Denmark caught fire (Solheim et al., 1992). Of the 498 passengers and crew aboard, 158 perished. The ferry did not sink. It was, in fact, towed to the Swedish port of Lysekil. All deaths were fire-related.

The Tanzanian ferry *Bukoba* capsized on Lake Victoria after leaving Bukoba and Kemondo Bay en route for Mwanza. The death toll in the accident on 21 May 1996 was over 500 (according to some estimates as high as 800 fatalities).

Several lessons can be learnt from the *Bukoba* sinking:

- According to reports from the scene, after a three-hour wait for equipment to arrive, efforts to extricate more than 500 trapped people only hastened the sinking.
- The ferry wavered before capsizing. A crew member reportedly gave more value to an 830-ton banana cargo than to human life. According to the report he responded to a suggestion of throwing cargo overboard by suggesting that people be forced off the overloaded ferry to allow stabilization.
- International press coverage was based virtually entirely on stories filed by local reporters. This determined the perspective and image conveyed to newspaper and radio audiences.
- The absence of an exact death toll is due to cultural norms that do not require Western-style registration, and an acceptance of "missing" as being a presumption of death.

The *Bukoba* was operating in open disregard of norms in the transportation business – without insurance, overloaded, without proper safety equipment, and without a proper emergency plan. Nor was the railroad company operating the ferry the only one to hold responsibility. An effective disaster response plan would have been quite difficult, paricularly since Tanzania has no search and rescue (SAR) team to respond. Perhaps most surprising, two years later the *East African* (Tanzanian weekly newspaper, 11–17 November 1998) reported that establishing an SAR team was still in the recommendation stages.

On 26 September 2000 the *Express Samina*, a ferry from Pireus en route to Paros, Greece hit rocks and sank before reaching port. Again there was no passenger list. According to the manifest, there were 447 passengers aboard in addition to the 64-member crew. This number of people, however, was only an estimate and did not include children under age five, since they are exempt from purchasing tickets.

This incident had a possible element of criminal negligence after it was reported that neither the captain nor his first officer was at the bridge at the time of the accident. The captain, Vassilis Yiannakisits, had been involved in two previous accidents; he was arrested and faced trial on criminal counts, as did two Greek coast guard officials.

The *Asia South Korea* ferry sank off the Philippines in the early morning hours of 23 December 1999, and the number of people aboard quickly became a key issue. The ferry was licensed to carry 614 passengers in addition to a crew of approximately 60 (i.e. about 674 persons). The company still denied that the ferry was overloaded, even when 711 passengers had been recovered safely in addition to the 27 bodies found. (Homicide charges were contemplated against the captain for allowing passenger overload.) Final numbers reached 699 survivors and 56 dead.

Ferry accidents in Philippine waters are endemic, and many of the problems remain uncorrected. Just as the *Asia South Korea* was overloaded in 1999, so was the *Dona Paz* overloaded (by 1300 passengers according to one report) when it collided with a tanker on 20 December 1987, killing 4341.[2]

The *Estonia* had undergone a safety inspection on 9 September 1994, then again on the day of its fatal last voyage. If the report of explosion is true, then the generally poor maintenance of the vessel assumes a different importance

[2] This was the most costly peacetime maritime accident in terms of loss of human life. The most costly wartime accident occurred on 30 January 1945, when at least 5348 people were killed when the German ship Wilhelm Gustloff, overloaded with German refugees, was torpedoed in the Baltic Sea near Poland by a Russian submarine.

The *Estonia* (built in 1979 by Meyer Werft as the *Viking Sally*), a three-times-a-week ferry from Tallin, Estonia to Stockholm, Sweden, sank in stormy seas off the Finnish coast at 0024 hours on 28 September 1994, causing the deaths of 1049 people. Water had entered the lower part of the ferry, first causing engine failure, then the ferry capsized. A report issued five years after the disaster determined the sinking to have been caused in part by explosion.

than originally thought. Shock and hypothermia caused most deaths. From a perspective of victimology, the cliché, "survival of the fittest," holds true. A disproportionate percentage of young males survived, as compared with women, the young and the elderly (Jackson, 1994).

CARGO SHIPS AND TANKERS

There are numerous accidents involving cargo ships and tankers. Most go unnoticed by the general public, because damage is relatively minor, most crews are rescued, and many of the incidents take place on the high seas far away from populated land areas. Incidents also usually involve specialized response teams, and not the standard municipal police, fire, and ambulance squads covered by the press. Only when there is some disruption of everyday life (such as an oil spill that reaches land) or an unusually high death count are these incidents reported by general media.

One advantage when dealing with cargo vessels is that the number and identity of crew members is known. That does not mean, however, that other persons are not aboard. On occasion there are passengers looking for inexpensive passage. One must also remember that not all passengers are necessarily legal. In October 2000, for example, 26 illegal emigrants from Fujian Province, China were found in a sealed container ready for shipment on a freighter to Long Beach, California. These emigrants were found alive, together with the food and drink they had prepared for the voyage. On 22 June 2000, however, 58 of 60 illegal immigrants to England were found dead in a truck that had crossed the English Channel aboard a ferry.

Rescue work is easier when the exact number of persons aboard a ship is definitely known. On 12 December 1999 at 0600 an emergency call was placed by the *Erika*, an oil tanker caught in a storm 50 kilometers off the coast of Brittany, France. The tanker was breaking in two, and rescue of the 26-member crew was requested. A marine patrol plane and a helicopter were dispatched to the scene. By 1100 the last crew member had been rescued. This relatively quick rescue was possible since the exact number of the crew was known.

When all members of the crew are accounted for after a disaster, it is only natural that emphasis moves from saving human lives to questions of ecology and animal life. It is necessary to consider as wide a range of hazards as possible. When the *Torrey Canyon* tanker leaked more than 50,000 tons of Gulf oil off the Scilly Isles (England) on 20 March 1967, it quickly became apparent that there was no local contingency plan for this type of emergency. This was also certainly true in the case of the *Erika,* in which cleaning took months, and an estimated 50,000 wild birds were killed from the pollution. Many of those dead birds had to be removed from the area.

The *Erika* is another example of two separate sites for a disaster. The bow and stern sank ten kilometers from each other, essentially causing a split recovery operation. The resultant oil spill was extensive with over 10,000 tons of heavy fuel oil (HFO) leaking from the bow. The viscosity of the fuel oil did not allow chemical dispersion.

Sometimes a ship breaks into two and one part stays afloat. That was the case with the *Volgoneft 248* tanker that broke apart near Istanbul in the early morning hours of 29 December 1999. The bow sank, but the stern remained afloat. All 17 crew members were rescued. The collision of the *Taishin Maru* and the *Yamato Maru* in the Chiba Prefecture of Japan on 12 January 2000 is an example in which the bow of the latter ship sank, and the stern sank.

The fate of the *Mineral Dampier* was not so fortunate. Twenty-seven sailors died (15 bodies retrieved over a search period conducted at different intervals during a 433-day period) on the cargo freighter on 22 June 1995. On that day the ship, of Belgian registry, collided with the *Hungin Madras* (South Korea) in the East China Sea at a point 400 km south of Nagasaki, Japan.

Although the latter vessel made the voyage safely to port, in the case of the *Mineral Dampier* structural damage (hence the true severity of the situation) went undetected until it was too late and the ship sank before making it to port.

It is not always a simple task to locate and positively identify a sunken ship. The *Mineral Dampier* (Kahana *et al.*, 1999) was located at a depth of 85 meters, using a small submarine hired by a commercial search contractor, Oceana Marine Research. Numerous typical problems were encountered in the operation; they can serve as an example of some of the issues involved even in a small maritime operation.

Over a 15-month period the salvage ship had to be re-moored several times to maintain position directly over the wreckage. On at least two occasions the air supply lines of divers became entangled in the wreckage. Water pressure significantly distorted voice contact with divers, requiring special "translation" for a severe "Donald Duck" effect. Search of the ship was difficult due to bent and broken metal, blocked passages, and lack of light. The final result of this situation is that it was not technically possible to retrieve the bodies of all deceased sailors.

Some bodies retrieved tended to flake when touched, necessitating a net so that pieces of skin would not be lost as the bodies were brought to the surface. Cold storage (–20°C) had to be arranged aboard ship before the bodies were transported to a land facility in Pusan, South Korea.

The accident occurred at an atmospheric temperature of 29°C and a surface water temperature of 20°–24°C. Conditions at the point where the ship rested were typically a 24-hour cyclical underwater current of 2.5 knots and temperatures ranging from 10° to 12°C. During the 433-day period of the retrieval operation the bodies found showed an increasing amount of damage, from "washerwoman's hands" on the second day to total skeletonization[3] on the last day.

[3] In the *Mineral Dampier* sinking Kahana *et al.* (1999) dismissed marine life as a factor in the skeletonization of three bodies. It is likely that the bodies underwent a flaking process as a result of water salts, and that skin and tissues were removed from the skeletons by water currents.

CONCLUSION

Even in the modern era with all of its advanced communications, disasters at sea can result in a large number of fatalities and in unique difficulties of response.

CHAPTER 27

LAND TRANSPORTATION

INTRODUCTION

Cars crash every day. Buses crash less often, but they do crash. These crashes are painful, and take a heavy emotional and physical toll.

Fires in vehicular accidents are more common that usually thought. According to the US Fire Administration (1999) fires in vehicles (primarily cars in highway accidents, but not only road vehicles) accounted for 24 per cent of reported fires and 17 per cent of fire-related deaths in the period 1987–1996.[1]

Notwithstanding such, these fires are very rarely of disaster proportions. The most common type of vehicular incident that does reach disaster proportions is when explosives are involved. (The Loma Prieta earthquake killed numerous people in vehicles, however this was not a transportation disaster; it was a natural disaster that affected a wide range of victims, including those in transit.)

There are different types of explosion incidents. Explosions can be classified as:

- Intentional – sabotage, terrorism, criminal.
- Unintentional – negligence.

There are also instances, both intentional and unintentional, in which an explosion is secondary to another incident, such as arson or accidental fire.

[1] These statistics carry the clear inference that cars have first aid kits with supplies specifically for fire/burn injuries.

CAR BOMBS

Car bombs have been a frequent method of terrorism in a number of countries. These bombs have been detonated in cars in various scenarios:

- door opening, motor ignition, or similar routine action (in murder scenarios the person operating the vehicle is unaware of the device);
- timer on parked or moving car;
- intentional detonation by person in the car after it has reached pre-designated point;

- detonation of vehicle by remote control from a distance (today cellular telephones have served as recipients of activating signals).

General defenses against car bombs are on-the-spot checking of license plates by inspectors (particularly illegally or "unusually" parked vehicles) and prohibition of parking in crowded public areas or in close proximity to sensitive buildings such as airline terminals.

Cars, sometimes laden with bombs, have been driven into crowds of people or into buildings. The placing of concrete barriers (sometimes with the more esthetic use as large plant holders) protects against both ramming and bombing.

If a car bomb does explode, there is often ensuing conflagration of the fuel system. From an investigative perspective it is important to retrieve information about the vehicle (make, model, serial number, license plate, etc.).

As an historical curiosity, on 16 September 1920 a blast on New York's Wall Street killed 34 people and injured some 300.[2] The explosion was traced to a wagon drawn by a roan horse (also killed). It took the police five days to identify the human dead, but efforts to identify the horse – for investigative purposes, of course – were unsuccessful. Those efforts included summoning an expert on personal identification and convening all of the area blacksmiths to find the person who made the horseshoes.

This is also an example of telephone lines overburdened and the convergence of an estimated 40,000 curiosity seekers at the scene.

[2] In a more recent example, on 30 January 2001 a wagon drawn by a donkey with a doll to imitate a driver was laden with explosives and used as a terrorist device in Israel. The wagon exploded before it reached its target. The only casualty was the donkey (Israel Television News).

Sometimes the effects of an explosion can be felt many miles away. Just as is the case with earthquakes, it is axiomatic that geological features are a major factor in the movement of energy along the surface of the earth.

RAIL COLLISION AND DERAILING

From a spectator perspective the most spectacular rail disaster is collision, and there are numerous examples of trains colliding, either into another train or into a vehicle at a rail crossing (either "official" or "improvised"). These events tend to generate media coverage, whether the collision is a "disaster" or an "accident."

One of the most serious train collisions in recent history occurred on 2 August 1999 at Gaisal, India, where 286 people died – reportedly when munitions exploded as the Brahmaputra Mail and the Awadh-Assam Express hit each other.

At the Amtrak crash in Maryland on 4 January 1987 disaster response was essentially divided into two segments, one on each side of what was a long multi-car train. At the Amtrak derailment in Iowa the accident scene was a quarter-mile long, with railroad cars tilted at a 45 degree angle.

An Amtrak train from Chicago to California derailed near Nowaday, Iowa (100 km from Omaha, Nebraska) on 18 March 2001. One person died, and some 90 people were injured (the great majority with only minor injuries).

A frequent problem in disaster planning and response is stereotyping. When one thinks of a train crash, the immediate reaction is to think of tracks on land. In a 1995 subway crash in New York the trains collided on a bridge, partially over land and partially over water. This meant a multi-faceted response that included precautions against people and debris falling both on land and into the water. In all some 200 passengers were evacuated (1 fatality, 63 injured); two responders suffered minor injury.

At 0612 on Monday, 5 June 1995 a Manhattan-bound "J" Line subway train collided with a Manhattan-bound "M" train on the Williamsburg Bridge that spans the East River, separating Brooklyn from Manhattan. Neither of the trains derailed. The Incident Command System was used for response administration.

The most frequent type of rail incident, however, is derailing. Although trains coming off the rail do include fatalities, it is the rare case that these incidents rarely reach disaster proportions, unless they are also accompanied by collision. When the derailments cause multiple deaths, newspaper coverage is extensive, and the impression is created that derailment is a primary danger to human life.

Another example of a non-stereotypical train derailment is the case of

Amtrak Train No. 2. The accident was a "worst case" scenario. The train derailed in dense fog on a railroad bridge in an area without vehicular access. A primary means of victim evacuation was by additional trains;[3] others were evacuated by boat and by helicopter.[4] It took responders some 18 minutes after the "May-Day" emergency message was received to determine the exact location of the derailment; the first fire-fighting equipment arrived more than an hour after the derailment. Not until morning did responders realize in what jurisdiction the accident had occurred, and who held primary responsibility. The number of possible victims was initially uncertain, since Amtrak kept no records of unticketed passengers (infants under age two).

[3] Victims were removed by train to a rear staging area from which they were taken by ambulance and bus (less seriously injured) to hospital.

[4] Landing in the area was not possible. The seven victims evacuated by helicopter were hoisted aboard.

> At 0253 on 22 September 1993 Amtrak Train No. 2, the Sunset Limited, en route from Los Angeles to Miami, derailed at Big Bayou Canot, Alabama (USA). Fire broke out, and two cars fell into a bayou (marsh). Forty-seven people (42 passengers and five train crew) were killed, 111 were injured, and 62 were uninjured (US Fire Administration, 1994).

The entire response operation showed very clearly that what was involved was more than a derailed train. One major concern during removal of the rail cars was a major communications fiber-optic cable that paralleled the tracks; the cable was pulled aside so that removal of the rail cars did not cause damage and loss of area communications.

> If Canadian statistics can be used as a general guideline (TSB, 1999), in 1998 there were 111 derailments in 1998 on Canadian railroads (not including rail yards and loading spurs). There were 99 rail-related deaths: 42 at rail crossings, and 56 as a result of trespassing. In other words, relatively few derailments caused fatalities.

Another type of rail accident involves hazardous materials (Hazmat). It should be stressed that these materials should not be dealt with until they are identified. Spraying water on some materials can increase, not lessen, danger. Hazmat are generally carried in the following types of rail cars:

- *Box car:* When entry through doors is not possible, roof is usually thinnest gauge steel. Some box cars are equipped with heaters for materials sensitive to temperature.

- *Flat car:* Several models exist (container, piggyback, depressed, etc.). These can carry Hazmat in railway containers, truck containers, or in independent packaging.
- *Gondola:* High sides and open roof. May contain Hazmat.
- *Hopper:* Cylinder car (open, covered, or pressurized) capable of carrying Hazmat.
- *Tanker:* Most common type of car carrying Hazmat. Leaks can be intentional (to relieve pressure) or accidental. (Source: *CP Rail System*)

Using Canadian statistics again (TSB, 1999), between 20–30 per cent of derailments include Hazmat cargo. Of the 244 derailments (including in-yard and on-spur) with Hazmat cargo involved, only in four cases was there release of hazardous material.

CONCLUSION

Even land transportation (cars, trucks, buses, trains, subways) is not totally safe. It is incumbent upon the disaster planner to recognize the hazards in each mode of transportation and identify response problems.

SECTION VI

CONCLUSION

CHAPTER 28

CONCLUSION

Transportation of all types has become an important and growing part of our daily lives. The entire nature of modern society is built around the supposition that people and goods can be moved rapidly from one place to another. As this book has shown, transportation is not without hazard. There is no doubt that contemporary society is increasingly vulnerable to transportation-related accidents and disasters. Our means of transportation expose more persons in great concentration under circumstances dependent on heavier, faster and more sophisticated mechanical and electronic devices. Although urban planning seeks to distance vehicles of all kinds from non-traveling populations, urban concentration does not often make such separation feasible. In many parts of the world it is even difficult to establish air corridors that do not pass over populated areas. No degree of precaution can make transportation fail-safe.

Many people have transportation risks imposed upon them without choice. Others voluntarily accept transportation-associated hazards. People are willing to undertake considerable risks when they are convinced of the advantages to be gained from their activities. Few would forgo the benefits of modern transportation to reduce its accompanying dangers. Even without proverbially disposing of the baby with the bath water, however, much can and must still be done to make twenty-first century transportation increasingly hazard-free.

Great strides have been made in making the machinery of transportation safer than it was in the past. Both mechanical and electronic equipment now have greater reliability than before due to improving durability of materials and more powerful computerized programs. Preventive maintenance and quality assurance practices are today widespread in all transportation-related industries.

People, however, are only marginally more sophisticated and capable than before, making the human factor in disaster all too prominent generally, and in transportation in particular. The human factor in planning, through operator error, to response, human limitations, mistakes and miscalculation, not to mention avarice and greed, plays a continuing role in making transportation disasters happen and making them worse when they do happen.

By now, a considerable literature exists concerning disasters around the world. In some areas that literature is more developed, and in other areas it is less developed. But, the literature does exist. At first glance each incident seems to have its own unique components. More systematic scrutiny, however, quickly reveals the surprising repetition of similar elements. Why then do we repeat the same costly mistakes? Why should we not at least learn from previous disasters to mitigate against recurrence of similar events? For one thing, we do not seem to document lessons as well as we describe events. For another, those who should take heed are not always as attentive as they ought to be.

Corner cutting, either motivated by pressure of time or cost, is often a factor. This can affect materials, stretched beyond their limits, or people, overworked or overloaded with responsibilities.

Ad hoc solutions, substitutions or changes of function are frequent culprits. So, too, are miscalculation and misjudgment, often made under the pressure to achieve immediate solutions. These may even be fueled by the over-confidence born of experience that demonstrates that nothing has ever gone wrong before.

It is not unusual to find that a rare combination of factors not usually found together is behind many disasters which, under normal circumstances, would not have occurred. Careful analysis of past experience can contribute to prevention or at least reduction of the price we pay for the odd disaster.

Finally, response needs careful consideration to forestall the errors of the past. One typical difficulty of such response is inter-agency and inter-departmental cooperation. Another is preparation for off-hours and inaccessible locations. Having a well thought-out and well-rehearsed disaster response in place, which is accepted by all participants, means that both human and material costs can be reduced. It requires professional leadership and represents a very worthwhile investment.

ADDENDUM: THE EVENTS OF 11 SEPTEMBER 2001

As this book was being prepared for press, terrorists hijacked four commercial airliners on 11 September 2001. Two of the planes struck the World Trade Center in New York, another plunged into the Pentagon in Arlington, Virginia, and the fourth crashed into an unpopulated area in rural Pennsylvania. In all, some 5000 people lost their lives. From every point of view these three tragedies fell under the rubric of terrorist-caused transportation disasters.

Chronology of Events on 11 September 2001

08:45 American Airlines Flight 011 Boeing 767 (Boston–Los Angeles carrying 81 passengers and 11 crew) crashed into the North Tower of the World Trade Center (WTC).

09:06 United Airlines Flight 175 Boeing 767 (Boston–Los Angeles carrying 56 passengers and 9 crew) crashed into the South Tower of the WTC.

09:40 American Airlines Flight 077 Boeing 757 (Washington/Dulles–Los Angeles carrying 58 passengers and 6 crew) crashed into the Pentagon.

10:00 South Tower of the WTC collapsed.

10:29 North Tower of the WTC collapsed.

10:37 United Airlines Flight 93 Boeing 757 (Newark–San Francisco carrying 38 passengers and 7 crew) crashed in Shanksville, Pennsylvania.

17:25 Another building at the WTC collapsed.

As of 21 October 2001, 4569 people were missing from the WTC, including those aboard the two hijacked jets, Flight 011 and Flight 175; 458 were confirmed dead. At the Pentagon 189 were believed dead, including those aboard Flight 077. In Pennsylvania 44 persons were killed aboard Flight 093.

Admittedly, this book does not advocate preparing for a disaster in the magnitude of the World Trade Center. One plans for the usual (and perhaps a little more). One cannot plan for the extremely large incident, even though history has just proven that it can (and did) happen. If there is sound disaster response programming in place, flexibility should allow its application even to the previously unthinkable.

Although the numbers of fatalities were numbing, the rules of response to transportation disaster held fast.

Now, only a month after the 11 September tragedies, is not the time to criticize planning or cast blame on aspects of the response. It is the time to carefully study that planning and response so that *constructive* lessons can be learnt for the future. That is a point that has been raised in this book – if post-incident analysis is transformed into a search for a villain, constructive lessons will never be learnt.

Those lessons will continue to be learnt over the coming months as just cleaning up the World Trade Center site continues. And as that work continues, both authors of this book recognize with admiration the work of all responders (both paid and volunteer) who came forth to help in time of need and tragedy.

BIBLIOGRAPHY

Aghababian, R.V. (1986), "Hospital Disaster Planning," *Topics in Emergency Medicine*, volume 7:4, pp. 46–54.

Alexander, D.A. (1991), "Psychiatric Intervention after the Piper Alpha Disaster," *Journal of the Royal Society of Medicine*, volume 84 (January), p. 8 ff.

Alexander, David (2000), "Scenario Methodology for Teaching Principles of Emergency Management," *Disaster Prevention and Management*, volume 9:2, pp. 89–97.

Almog, Joseph and Glattstein, Baruch (1997), "Detection of Firearms Imprints on Hands of Suspects: Study of the PDT-based Field Test, *Journal of Forensic Sciences*, 42:6, November, pp. 993–996.

Almog, Joseph and Levinson, Jay (2000) "Forensic Science: A Tool in the Fight against Terrorism," *Police Chief*, 67:10, pp. 131–136.

American Board of Forensic Odontology (1995), "Body Identification Guidelines," *Journal of the American Dental Association*, volume 125 (September), pp. 1244–1254.

American Psychiatric Association (1980), *Diagnostic and Statistical Manual of Mental Disorders*, 3rd edition III. Washington, DC.

American Psychiatric Association (1987), *Diagnostic and Statistical Manual of Mental Disorders*, 3rd edition, revised. Washington, DC.

American Psychiatric Association (1994), *Diagnostic and Statistical Manual of Mental Disorders*, 4th edition. Washington, DC.

Amoedo, Oscar (1897), "The Role of the Dentists in the Identification of the Victims of the Catastrophe of the 'Bazar de la Charit,' Paris 4 May 1897," *Dental Cosmos*, volume 39, pp. 905–912.

Amoedo, Oscar (1898), *L'Art Dentaire en Medicine Legale*. Paris: Masson et Cie Editeurs. Libraires de l'Academie de Medecine.

auf der Heide, Erik (1989), *Disaster Response: Principles of Preparation and Coordination*. St Louis: CV Mosby Company.

Averch, Harvey and Dluhy, Milan J. (1997), "Crisis Decision Making and Management," in Peacock, Walter Gillis, Morrow, Betty Hearn, and Gladwin, Hugh, *Hurricane Andrew: Ethnicity, Gender, and the Sociology of Disasters*. Miami: International Hurrican Center.

Baldi, J.J. (1974), "Project Research: Anatomy of a Survey under Disaster Conditions," *Gerontologist*, volume 14, pp. 100–105.

Barbash, Gabriel I., Yoeli, Naomi, Ruskin, Stephen M. and Moeller, Dade W. (1986), "Airport Preparedness for Mass Disaster: A Proposed Schematic Plan," *Aviation, Space and Environmental Medicine*, volume 57, pp. 77–81.

Beaton, Randall D. and Murphy, Shirley A. (1995), "Working with People in Crisis: Research Implications," in C.R. Figley (ed.), *Compassion Fatigue: Coping with Secondary Traumatic Stress Disorder in Those Who Treat the Traumatized*. New York: Brunner/Mazel Publishers, pp. 51–81.

Benjamin/Clark Associates (1984), *Fire Deaths: Causes and Strategies for Control*. Lancaster, Pennsylvania: Technomic Publishing Company, Inc.

Blashan, Sue and Quarantelli, E.L. (n.d.), "From Dead Body to Person: The Handling of Fatal Mass Casualties in Disasters," Preliminary Paper No. 61, Columbus, Ohio: Disaster Research Center, Ohio State University.

Brooks, Carol Carlsen (1986), "The Day It Rained Death from the Sky," *American Fire Journal* (October).

Brown, C. (1995), *Chaos and Catastrophe Theories*. Thousand Oaks, CA: Sage Publishing.

Brown, Carl E. (1999), "Remote Sensing Aircraft Play a Vital Role," in *Spill Technology Newsletter* (Environment Canada), volume 24:3–4.

Brown, Kenneth A. (1984), "Dental Identification of Unknown Bodies," *Annals of the Academy of Medicine*, volume 13:1 (January), pp. 3–7.

Brown, Kenneth A. and Powell, Graham L.P. (1975a), *Supportive Evidence in Dental Identifications*, paper presented at Fourth Australian National Symposium on Forensic Science, Perth, Australia.

Brown, Kenneth A. and Powell, Graham L. P. (1975b), *Forensic Science Teamwork in Dental Identifications*, Paper presented at Seventh International Meeting of Forensic Sciences, Zürich, Switzerland.

Brown, T. (1952), "Medical Identification in the Noronic Disaster," *Proceedings of the American Academy of Forensic Sciences*, volume 2:1, pp. 109–119.

Bruggeman, Lt. Co. (1992), "The Zeebrugge Ferry Disaster: Introduction and Background to the DVI Organization," *Interpol: International Criminal Police Review*, Number 437–438, pp. 6–8.

Busuttil, A. and Jones, J.S.P. (1992), "The Certification and Disposal of the Dead in Major Disasters," *Medicine, Science and the Law*, volume 32:1, pp. 9–13.

Butcher, J.N. and Hatcher, C. (1988), "The Neglected Entity in Air Disaster Planning – Psychological Services," *American Psychologist*, volume 433:9, pp. 724–729.

Cameron, J.M. and Sims, Bernard G. (1973), *Forensic Dentistry*. Edinburgh: Churchill Livingstone.

Cannon, Walter (1914), "The Interrelation of Emotions as Suggested by Recent Physiological Researches," *American Journal of Psychology*, volume 25, p. 256 ff.

Chapple, Emery W. Jr. (1972), "Air Disaster Recovery Operations in Remote Areas," Washington, DC: *FBI Law Enforcement Bulletin*, June.

Cheu, D.H., Hays, M.B. and Stepanki, J.X. (1976), "Planning an Airport Disaster Drill," *Aviation, Space and Environmental Medicine*, volume 47:5, pp. 556–560.

Clark, Derek H. (1986), "Dental Identification Problems in the Abu Dhabi Air Accident," *American Journal of Forensic Medicine and Pathology*, volume 7:4, pp. 317–321.

Clark, Derek H. (1991), "Dental Identification in the Piper Alpha Oil Rig Disaster," *Journal of Forensic Odonto-Stomatology*, volume 9:2, pp. 37–45.

Clark, Derek H. (ed.) (1992), *Practical Forensic Odontology*. Oxford: Wright (Butterworth-Heinemann Ltd).

Clark, Michael A., Clark, Stanley R. and Perkins, David G. (1989), "Mass Fatality Aircraft Disaster Processing," in *Aviation, Space and Environmental Medicine*, (July), pp. A64–A73.

Coglianese, Cary and Howard, Margaret (1998), "Getting the Message Out: Regulatory Policy and the Press," *Harvard International Journal of Press/Politics*, volume 3.

Collins, R. and Leathley, B. (1995), "Psychological predispositions to Errors," in *Safety, Reliability and Failure Analysis. Safety and Reliability*, volume 14:3, pp. 6–42.

Corey, G. (1991), *Theory and Practice of Counseling Psychotherapy*. Belmont, California: Brooks Cole.

Covey, Stephen R., Merrill, A. Roger and Merrill, Rebecca R. (1994), *First Things First: To Live, to Love, to Learn, to Leave a Legacy*. New York: Simon & Schuster.

Cullen, S.A. and Turk, E.P. (1980), "The Value of Postmortem Examination of Passengers in Fatal Aviation Accidents," in *Aviation, Space and Environmental Medicine*, pp. 1071–1073.

Cuny, Frederick (1983), *Disasters and Development.* New York: Oxford University Press.

de Goyet, Clause de Ville (1999), "Stop Propagating Disaster Myths," *Australian Journal of Emergency Management,* volume 14:4, pp. 26–28.

DeAtley, Craig A. (1982), "One Hundred Fifty Minutes: A Double Tragedy in the Nation's Capital," *Journal of Emergency Medical Services,* March, pp. 26–33.

Doyle, C.T. and Bolster, M.A. (1992), "The Medico-legal Organization of a Mass Disaster – the Air India Crash 1985," *Medicine, Science and the Law,* volume 32:1, pp. 5–8.

Drabek, Thomas (2001), "Understanding Employee Responses to Disaster," *Australian Journal of Emergency Management,* volume 15:4 (Summer), pp. 15–21.

Drory, Margalit, Posen, Jennie, Vilner, Doron and Ginzburg, Karni (1998), "Mass Casualties: An Organizational Model of a Hospital Information Center in Tel Aviv," in *Social Work in Health Care,* volume 27:4, pp. 83–94.

Dunn, B.W. (1917), *Report on Explosion at Black Tom, New Jersey, 30 July 1916.* New York: Bureau of Explosives.

Durant, Henry (1862, reprint 1986), *A Memory of Solferino.* Geneva: International Committee of the Red Cross.

Dynes, Richard R. (1974), *Organized Behavior in Disaster.* Columbus, Ohio: Disaster Research Center.

Dynes, Russell R. (1994), "Community Emergency Planning: False Assumptions and Inappropriate Analogies," *International Journal of Mass Emergencies and Disasters,* volume 12:2, pp. 141–158.

Engel, G. (1964), "Grief and Grieving," *American Journal of Nursing,* volume 64, pp. 93–98.

Emson, E.H. (1978), "Problems in the Identification of Burn Victims," *Canadian Forensic Science Journal,* volume 11:3, pp. 229–236.

Eyre, A. (Spring 1999) In Remembrance: Post-Disaster rituals and symbols. *Australian Journal of Emergency Management,* volume 14:3, pp. 23–29.

Farberow, Norman L. and Frederick, Calvin J. (1978; reprinted 1983, 1986), *Training Manual for Human Service Workers in Major Disasters.* Rockville, Maryland: National Institute of Mental Health, U.S. Department of Health and Human Services.

Fatteh, Abdullah (1973), *Handbook of Forensic Pathology.* Philadelphia, Pennsylvania: J.B. Lippincott Company.

Federal Bureau of Identification (1949, revised 1968), "Problems and Practices in Fingerprinting the Dead," *FBI Law Enforcement Bulletin,* April 1949/November 1968.

Feist, Andy (n.d.), *The Effective Use of the Media in Serious Crime Investigations.* UK Home Office: Policing and Reducing Crime Unit, Paper 120.

Ferrier, Norman (1999–2000), "Demographics and Emergency Management: Knowing Your Stakeholders," *Australian Journal of Emergency Management,* volume 14:4, pp. 2–4 (Summer issue).

Figley, Charles R. (1995), Compassion Fatigue as Secondary Stress Disorder: An Overview," in C.R. Figley (ed.), *Compassion Fatigue: Coping with Secondary Traumatic Stress Disorder in Those Who Treat the Traumatized.* New York, Brunner/Mazel Publishers, pp. 1–20.

Fisher, Barry J. (1993), *Techniques of Crime Scene Investigation,* 5th edition. Boca Raton, Florida: CRC Press.

Fodor, Joe (1999), "The Crash of Flight 826," *Brooklyn Bridge,* volume 4:12 (November–December), pp. 58–65.

Frederick, Calvin J. (n.d.), *Aircraft Accidents: Emergency Mental Health Problems.* Rockville, Maryland: National Institute of Mental Health.

Freudenberger, H.J. (1974), "Staff Burnout," *Journal of Social Issues,* volume 30:1, pp. 159–165.

Friedsam, H.J. (1962), "Older Persons in Disaster," in Baker, G.W. and Chapman, D.W. (eds), *Man and Society in Disaster.* New York: Basic Books, pp. 151–184.

Fruin, John J. (1981), "Causes and Prevention of Crowd Disasters," in *Student Activities* (October), pp. 48–53.

Garner, Ana C. (1996), "Reconstructing Reality: Interpreting the Aeroplane Disaster News Story," in *Disaster Prevention and Management*, volume 5:3, pp. 5–15.

Gerber, Samuel R. (1952), "Identification in Mass Disasters by Analysis and Correlation of Medical Findings," *Proceedings of the American Academy of Forensic Sciences*, volume 2:1, pp. 82–98.

Geide, Bernd (1992), "IDKO Action in Thailand," *Interpol: International Criminal Police Review*, Number 437–438, pp. 29–32.

Gilchrist, Jim (1992), "The Lockerbie Air Disaster or the Shortest Day," *Interpol: International Criminal Police Review*, Number 437–438, pp. 23–28.

Glazer, Howard (1996), "CAV-ID," *Disaster Prevention and Management*, volume 5:3, p. 51.

Gleik, J. (1997), *Chaos*. London: Heinemann.

Glick, I.O., Weiss, R.S. and Parkes, C.M. (1974), *The First Year of Bereavement*. New York: Wiley & Sons.

Granger, K.J. and Johnson, R.W. (1994), *Hazard Management: Better Information for the 21st Century*. Canberra: Emergency Management Australia.

Granot, Hayim (1993), "Anti-Social Behaviour Reconsidered," in *Disaster Management*, volume 5:3, pp. 142–144.

Granot, Hayim (1996), "Disaster Subcultures," in *Disaster Prevention and Management*, volume 5:4, pp. 36–40.

Granot, Hayim (1996a), "The Impact of Disaster on Mental Health," in *Counseling*, (May), pp.140–143.

Granot, Hayim (1998), "The Human Factor in Industrial Disaster," in *Disaster Prevention and Management*, 7:2, pp. 92–102.

Granot, Hayim (1999), "Emergency Inter-organizational Relationships," in *Disaster Prevention and Management*, volume 8:1, pp. 21–26.

Granot, Hayim and Levinson, Jay (2001), *Terrorist Bombing: The New Urban Threat*, Tel Aviv (Israel): Dekel Publications.

Gray, Charles and Knabe, Harold (1981), "The Night the Skywalks Fell," *Firehouse*, September, pp. 66–70, 132–133.

Grayson, David (1988), *Terror in the Skies: The Inside Story of the World's Worst Air Crashes*. Secaucus, New Jersey: Citadel Press.

Great Britain Home Office (n.d.), *Dealing with Disaster*, 3rd edition. London: Home Office. ISBN 185 893 9208

Greenberg, Reuben M., Wiley, Charles, Youngblood, Glenn, Wetsell, Herbert and Doyle, H. (1990), "The Lessons of Hurricane Hugo: Law Enforcement Responds," *Police Chief*, volume 57:9, pp. 26–33.

Guerin, Michael (1990), "Disaster Operations: Not Business as Usual," *FBI Law Enforcement Bulletin*, volume 59:12, pp. 1–6.

Haertig, A. (1990), "Identification des victimes de grande catastrophe," *Urgences*, volume 9, pp. 392–395.

Harvey, Warren (ed.) (1976), *Dental Identification and Forensic Odontology*. London: Henry Kimpton Publishers.

Hazen, Robert J. and Phillips, Clarence E. (1982), *Field Disaster Identification: Preparation, Organization, Procedures*. Quantico, Virginia: FBI Academy Forensic Science Training Unit.

Hershiser, Marvin R. and Quarantelli, E.L. (1976), "The Handling of the Dead in a Disaster," *Omega: Journal of Death and Dying*, volume 7:3, pp. 195–208.

Hinkes, Madeleine J. (1989), "The Role of Forensic Anthropology in Mass Disaster Resolution," *Aviation, Space, and Environmental Medicine*, volume 60:7 (suppl.), pp. A60–A63.

Hodgkinson, P.E. and Stewart, M. (1988), "Missing, Presumed Dead," *Disaster Management*, volume 1:1, pp. 11–14.

Hooft, Peter J., Noji, Eric K. and van de Voorde, Herman P. (1989), "Fatality Management in Mass Casualty Incidents," *Forensic Science International*, volume 40:1, pp. 3–14.

Houts, Peter S. and Goldhaber, Marylin (1981), "Mobility of the population within 5 miles of Three Mile Island during the period from August 1979 through July 1980." Report / submitted to the TMI Advisory Panel on Health Research Studies of the Pennsylvania Department of Health. Harrisburg, Pennsylvania: Dept. of Health.

Howard, B.W. (1999), "Managing Volunteers," *Australian Journal of Emergency Management*, volume 14:3 (Spring), pp. 37–39.

Hutton, Janice R. (1976), "The Differential Distribution of Death and Disaster: A Test of Theoretical Propositions," *Mass Emergencies*, volume 1, pp. 261–266.

Institute of Civil Defense and Disaster Studies (2000), "NCCP Proposals for a New Civil Protection Act," *Alert*, January–March, p. 4.

Jackson, A.A. (1988), *The Institute of Civil Defence 1938–1955: A Short History of Its First Seventeen Years*, London: Institute of Civil Defence.

Jackson, James O. (1994), "The Cruel Sea," *Time* (European Edition), 10 October, pp. 16–21.

Jain, M.C. (1997), *Jain Committee of Inquiry Interim Report*. New Delhi, India.

Jeffreys, A.J.; Wilson, V. and Thein, S.L. (1985), "Individual-Specific 'Fingerprints' of Human DNA," *Nature*, volume 316, pp. 76–79.

Jones, David R. (1985), "Secondary Disaster Victims: The Emotional Effects of Recovering and Identifying Human Remains," *American Journal of Psychiatry*, volume 242:3, pp. 303–307.

Kahana, Tzippy; Almog, Joseph; Levy, Joseph; Shmeltzer, Elie; Spier, Y.; Hiss, Jehudah (1999), "Marine Taphonomy: Adipocere Formation in a Series of Bodies Recovered from a Single Shipwreck," *Journal of Forensic Sciences*, volume 44:5, pp. 897–901.

Kahill, S. (1988), "Interventions for Burnout in the Helping Professions: A Review of the Empirical Evidence," *Canadian Journal of Counseling Review*, volume 22:3, pp. 310–342.

Kanarev, Nicholas (2001), "Assessing the Legal Liabilities of Emergencies," in *Australian Journal of Emergency Management*, volume 16:1, pp. 18–22.

Kaplan, Meier and Almog, Joseph (1983), "Field Diagnostic Examinations for Forensic Purposes." *Police Chief*, September, pp.30–33.

Keiser-Nielsen, S., Johanson, G. and Solheim, Tore (1981), "The Dental X-Ray File of Crew Members in the Scandinavian Airline System (SAS)," *Aviation, Space, and Environmental Medicine*, volume 52:11, pp. 691–695.

Kerr, Joanna (2001), "Swissair 111 – Aftermath of a Tragedy: Canada's Largest DNA Effort Offers Lessons for the Future," *RCMP Gazette*, volume 63:2, pp. 21–22.

Kletz, Trevor A. (1996a), "Disaster Prevention: Current Topics," in *Disaster Prevention and Management*, volume 5:2, pp. 36–41.

Kletz, Trevor A. (1996b), "Two Views – The Public's and the Experts'," in *Disaster Prevention and Management*, volume 5:4, pp. 41–46.

Kübler-Ross, Elisabeth (1969 and reprints), *On Death and Dying*. New York: Macmillan & Company.

Ladkin, Peter (1996), *Aircraft Accident Information Report 96-5: China Airlines Airbus Industrie A300B4-622R, B1816 Nagoya Airport, April 26, 1994*. Tokyo: Ministry of Transport, Aircraft Accident Investigation Commission.

Lawrence, Susan C. (1998), "Beyond the Grave – The Use and Meaning of Human Body Parts: A Historical Introduction," in Weir, Robert F. (ed.), *Stored Tissue Samples: Ethical, Legal and Public Policy Implications*. Iowa City: University of Iowa Press.

Le Bon, Gustave (1897), *The Crowd: A Study of the Popular Mind*. London: T.F. Unwin.

Ledger, Don (2000), *Swissair Down: A Pilot's View of the Crash at Peggs Cove*. Halifax, Nova Scotia: Nimbus Publishing.

Levinsohn, Alexander (1999), *Aspects of Disaster Victim Identification, Part I: Recognition and Marks*. Jerusalem: Israel Police Rabbinate.

Levinson, Jay (1993), "Contingency Planning for an Airport Crash," *Airport Forum*, February, pp. 48–50.

Levinson, Jay (1993a) "When Your Business Employees are Killed in an Air Crash," *Disaster Recovery* (USA), January–March.

Levinson, Jay (1998a). "Coverage of the Israeli Press: The Machane Yehudah Bombing," *Disaster Prevention and Management*, volume 3.

Levinson, Jay (1998b). "The Day After the Ben Yehudah Street Bomb: Israeli Press Coverage," *Disaster Prevention and Management*, volume 3.

Levinson, Jay and Amar, Shimon (1999), "Disaster Response Kits," *Disaster Prevention and Management*, volume 8:4, pp. 277–279.

Levinson, Jay and Shmeltzer, Elie (1992), "The Crash of Bus 405," Lyons, France: *Interpol International Criminal Police Review*, July–October, pp. 21–22.

Lindbloom, Roland E. (1950), "TNT on the Loose," Newark (New Jersey) *Sunday News* (2 April).

Lindbloom, Roland E. (1950a), "So Red the Fire," Newark (New Jersey) *Sunday News* (19 March), pp. 17–20.

Lindell, Michael K. and Perry, Ronald W. (1992), *Behavioral Foundations of Community Emergency Planning*. Washington, DC: Hemisphere Publishing Corporation.

Lindemann, Erich (1944), "Symptomology and Management of Acute Grief," *American Journal of Psychology*, volume 101, pp. 141–148.

Lorton, L. and Langley, W. (1984), *Postmortem Identification: A Computer Assisted System*, Washington, DC: US Army Institute of Dental Research.

Lorton, L. and Langley, W. (1986), "Design and Use of a Computer Assisted Postmortem Identification System," *Journal of Forensic Science*, volume 31, pp. 972–981.

Luntz, Lester (1997), "History of Forensic Dentistry," *Dental Clinics of North America*, volume 21:1, pp. 7–17.

Luntz, Lester L. and Luntz, Phyllys (1973), *Handbook for Dental Identification Techniques in Forensic Dentistry*, Philadelphia: J.B. Lippincott Company.

Maher, George F. (1990), "The Tragedy of Avianca Flight 052," *Police Chief*, volume 57:9 (September), pp. 39–43.

Mann, Robert W. and Ubelaker, Douglas H. (1990), "The Forensic Anthropologist," *FBI Law Enforcement Bulletin*, volume 59:7, pp. 20–26.

Maslach, C. and Jackson, S.W. (1981a), "Measurement of Experienced Burnout," *Journal of Occupational Behavior*, volume 2, pp. 99–113.

Maslach, C. and Jackson, S.W. (1981b), *The Maslach Burnout Inventory*, Research Edition. Palo Alto, California: Consulting Psychologists Press, Inc.

McGee, Rod (2001), "Tasman Bridge Disaster: 25th Anniversary Memorial Service," *Australian Journal of Emergency Management*, volume 15:4 (Summer), pp. 10–14.

Meade, Peter W. (1990), The Avianca Crash: Mission Accomplished," *Fire Command*, June 1990, pp. 24–28.

Michaels, Mark (1990), "Earthquake: Personal Accounts by Aviators of Their Roles during and after the San Francisco Earthquake," *Air Beat*, March–April, pp. 22–29.

Mileti, Dennis S. (1999), *Disasters by Design: A Reassessment of Natural Hazards in the United States*. Washington, DC: John Henry Press.

Mitchell, Jeffrey T. and Bray, G. (1990), *Emergency Services Stress*. Englewood Cliffs, New Jersey: Prentice Hall.

Mitchell, Jeffrey T. and Everly, George S. (1996), *Critical Incident Stress Debriefing: An Operations Manual for the Prevention of Traumatic Stress Among Emergency Service and Disaster Workers*. Ellicott City, Maryland: Chevron Publishing Corporation.

Mittleman, Roger E., Davis, Joseph H., Kasztl, Warren and Graves, Wallace M. Jr (1992), "Practical Approach to Investigative Ethics and Religious Objections to the Autopsy," *Journal of Forensic Sciences*, volume 37:3, pp. 824–829.

Moody, Don (1998), *America's Worst Train Disaster: The 1910 Wellington Tragedy*. Plano, Texas: Abique.

Morris, Charles (1906/1986) *The San Francisco Calamity by Earthquake and Fire . . . Told by Eye Witnesses, Including Graphic and Reliable Accounts of all Great Earthquakes and Volcanic Eruptions in the World's History, and Scientific Explanations of their Causes.* Philadelphia, Chicago [etc.]: The J. C. Winston Co.

Morrow, Betty Hearn and Peacock, Walter Gillis (1997), "Disaster and Social Change," in Peacock, Walter Gillis, Morrow, Betty Hearn and Gladwin, Hugh, *Hurricane Andrew: Ethnicity, Gender, and the Sociology of Disasters.* Miami: International Hurrican Center.

Muir, E. (1935), "Chiropody in Crime Detection," *The Chiropodist,* volume 22, pp. 165–166.

Nassau County (New York) (1995), *Disaster Control Plan: Limited Area Disasters – Police and Fire Response (including Emergency Medical Services),* 33 pages. Nassau County, NY Police.

Nepal Accident Investigation Commission (1993), *Aviation Accident Report: Thai Airways International Ltd., Airbus Industrie A310–304, HS-BID, Near Kathmandu, Nepal 23NM NNE, 31 July 1992.* Kathmandu.

O'Connor, John J. (1987), *Practical Fire and Arson Investigation.* (Series: Practical Aspects of Criminal and Forensic Investigations). New York: Elsevier.

Okerby, Peter (2001), "Evacuation of a Passenger Ship – is Panic a Major Facor?," *Australian Journal of Emergency Management,* volume 10:1, Autumn, pp. 8–14.

Osterburg, James W. and Ward, Richard H. (1992), *Criminal Investigation: A Method for Reconstructing the Past.* Cincinnati, Ohio: Anderson Publishing Company.

Pearce, Kathy *et al.* (1985), "Study of PTSD in Vietnam Veterans," *Journal of Clinical Psychology,* volume 41:1, pp. 9–14.

Pearson-Fry, Brian (2000), "Aviation Security: A Systems Approach," *International Security Review,* volume 115, pp. 15–17.

Perrier, D.C. and Toner, R. (1984), "Police Stress: The Hidden Foe," *Canadian Police College,* volume 8:1, pp. 15–26.

Perry, Ronald W. (1995), "The Structure and Function of Community Emergency Operations Centres," *Disaster Prevention and Management,* volume 4:5, pp. 37–41.

Perry, Ronald W. and Lindell, Michael K. (1996), "Aged Citizens in the Warning Stages of Disaster: Reexamining the Evidence," *International Journal of Aging and Human Development,* volume 44, pp. 300–311.

Perry, Walt L., Schrader, John Y. and Wilson, Barry M. (1993), *A Viable Resolution Approach to Modeling Command and Control in Disaster Relief Operations,* Santa Monica, California: Rand Corporation.

Peschel, Richard E. and Peschel, Enid Rhodes (1983). "Ritual and the Death Certificate: Case Histories, Literary Histories," *The Pharos,* Spring 1983, pp. 11–16.

Poulshock, S.W. and Cohen, E.S. (1975), "The Elderly in the Aftermath of a Disaster," *Gerontologist,* volume 15, pp. 357–361.

Poulton, Geoff (1988), *Managing Voluntary Organizations.* New York: John Wiley & Sons.

Price, Samuel Henry (1920), *Catastrophe and Social Change Based upon a Sociological Study of the Halifax Disaster.* New York: Columbia University Faculty of Political Science.

Powell, J.W. and Rayner, Janette (1952), *Progress Notes: Disaster investigation, July 1, 1951 – June 30, 1952.* Edgewood, Maryland: US Army Chemical Center, Chemical Corps Medical Laboratories.

Quarantelli, E.L. and Dynes, Russell R. (1970), "Property Norms and Looting: Their Patterns in Community Crises," *Phylon,* volume 31:2, pp. 168–182.

Quarantelli, E.L. (1979), *Studies in Disaster Response and Planning.* Newark, Delaware: University of Delaware Disaster Research Center.

Quarantelli, E.L. (1996), "Local Mass Media Operations in Disasters in the USA." *Disaster Prevention and Management,* volume 5:5, pp. 5–10.

Rando, Therese A. (1984). *Grief, Dying and Death: Clinical Interventions for Caregivers.* Champaign, Illinois: Research Press Company.

Reinholtd, Samantha (1999–2000), "Managing Change within the Emergency Services to Ensure the Long-term Viability of Volunteerism," *Australian Journal of Emergency Management*, volume 14:4, pp. 6–9.

Richardson, Gary (1996). "So What Have We Got Here?" in Kuban, Ron (ed.), *Canadian Fire Officer's Guide to Emergency Management*. Calgary: Pendragon Publishing, Ltd.

Roemer, L. and Borkovec, T. (1994), "Effects of Suppressing Thoughts about Emotional Material," *Journal of Abnormal Psychology*, volume 103, pp. 467–474.

Rogers, Everett M and Berndt, Matthew R. (1992), "Mass Media Estimates of a Disaster's Severity: The 1989 Loma Prieta Earthquake," *Disaster Management*, volume 4:2.

Salomone, Jeffrey III, Sohn, Anton, P., Ritzlin, Roger, Gauthier, Joseph, H. and McCarty, Vernon (1987), "Correlations of Injury, Toxicology, and Cause of Death to Galaxy Flight 203 Crash Site," *Journal of Forensic Sciences*, volume 32:5, pp. 1403–1415.

Scanlon, Joseph (1991), "Reaching Out: Getting the Community Involved in Preparedness," *Canadian Police College Journal*, volume 15:1, p. 39 ff.

Scanlon, Joseph (1992), "Not Just Bigger but Different: The Problems of Planning for Disaster," *RCMP Gazette*, volume 54:6, pp. 1–10.

Selye, Hans (1979), *The Stress of My Life*. New York: Van Nostrand.

Shane, Curtis M. (1990), "Earthquake: 15 Seconds of Terror," Paper presented at International Symposium on the Forensic Aspects of Mass Disasters and Crime Scene Reconstruction, 23–29 June.

Shaw, Richard (2001), "Don't Panic: Behaviour in Major Incidents," *Disaster Prevention and Management*, volume 10:1, pp. 5–10.

Siegel, Robert, Sperber, Norman D. and Trieglaff, Ario (1977), "Identification through the Computerization of Dental Records," *Journal of Forensic Sciences*, volume 22:23, pp. 434–442.

Silverstein, Martin E. (1992), *Disasters: Your Right to Survive*. McClean, Virginia: Brassey's (US), Inc.

Sime, Jonathan D. (1980), "The Concept of 'Panic'," Chapter 5 in Canter, David (ed.), *Fires and Human Behaviour*. Chichester (U.K.): John Wiley & Sons.

Skertchly, Allan and Skertchly, Kristen (2001), "Catastrophe Management: Coping with Totally Unexpected Extreme Disasters," *Australian Journal of Disaster Management*, volume 16:1, pp. 23–33.

Sloan, Harry S. (1995), "A Mid-Sized Department's Identification Response to Mass Disaster," *Journal of Forensic Identification*, volume 45:3, pp. 275–279.

Smale, Sydney (2000), "The Port Arthur Anniversary Services: Post-Disaster Rituals and Symbols," *Australian Journal of Emergency Management*, volume 15:1, pp. 2–5.

Smith, Brion C., Fisher, Deborah L., Weedn, Victor W., Warnock, Gary R. and Holland, Mitchell M. (1993), "A Systematic Approach to the Sampling of Dental DNA," *Journal of Forensic Sciences*, volume 38:5, pp. 1194–1209.

Smith, Denis (1992), "The Kegworth Air Crash: A Crisis in Three Phases?" in *Disaster Management*, volume 4:2, pp. 63–72.

Solheim, Tore, Lorentsen, Magne, Sundnes, Per Kristian, Bang, Gisle and Bremnes, Lasse (1992), "The 'Scandanavian Star' Ferry Disaster 1990 – A Challenge to Forensic Odontology," *International Journal of Legal Medicine*, 1992, pp. 617–623.

Solheim, Tore, Rønning, Steinar, Hars, Bjørn and Sundnes, Per Kristian (1982), "A New System for Computer Aided Dental Identification in Mass Disasters," *Forensic Science International*, volume 20, pp. 127–131.

Stanton, P., Jarvis, R. and Deligianni, E. (2000), "The Duzce, Turkey Earthquake," Institute of Civil Defense and Disaster Studies: *Alert*, January–March, pp. 18–19.

Stark, Andrew M. (1990), "The 1989 California Earthquake: Cypress Structure Rescue Operations," unpublished paper.

Stevens, Peter J. (1970), *Fatal Civil Aircraft Accidents: Their Medical and Pathological Investigation*. Bristol (UK): John Wright & Sons Ltd.

Stevenson L, and Hayman M. (1981), Local Government Disaster Protection: Final technical report. International City Management Association, Washington, DC, 1981.
Stout, Jack and Smith, Patrick (1981), "Nightmare in Kansas City," JEMS, September, pp. 34–45.
Stuart, Peter (1985), *Mass Victim Identification*. London: Metropolitan Police Form 3140, Airport District Training Section.
Taft, Brian and Reynolds, Simon (1994), *Learning from Disasters*. Oxford: Butterworth-Heinemann, Ltd.
Taylor, A.J.W. (1968), "A Search among Borstal Girls for Psychological and Social Significances of their Tattoos," *British Journal of Criminology*, volume 8, p. 170 ff.
Taylor, A.J.W. and Frazer, A.G. (1980), "Interim Report of the Stress Effects on the Recovery Teams after the Mount Erebus Disaster, November 1979," *New Zealand Medical Journal*, No. 91, pp. 311–312.
Tenhunen, Matti and Makel, Hanu (1993), "The Finnish Computer Assisted Disaster Victim Identification System," unpublished paper.
Thompson, Robert L., Manders, William W. and Cowan, William R. (1987), "Postmortem Findings of the Victims of the Jonestown Tragedy," *Journal of Forensic Sciences*, volume 32:2 (March), pp. 433–443.
Transportation Safety Board (TSB) (1999), *Annual Report to Parliament 1998–1999*. Hull, Québec, Canada: Minister of Public Works and Government Services.
Tumelty, David (1990), *Social Work in the Wake of Disaster*. London: Jessica Kingsley Publishers.
Turner, B.A. (1992), "Crowd Behaviour: Lessons to be learned from Summerland," in *Easingwood Papers No. 4: Lessons Learned from Crowd-related Disasters*. Easingwook, England: Emergency Planning College.
US Defense Civil Preparedness Agency (1976), *Management of Medical Problems from Population Relocation*. Washington, DC: US Department of Defense.
US Department of Army (1968), *Civil Disturbances and Disasters*. Washington, DC: United States Government Printing Office (GPO).
US Federal Disaster Assistance Administration (1978), *Documenting Disaster Damage*. Washington, D.C.: US Department of Housing and Urban Development.
US Fire Administration (1987), *Technical Report Series: Ramada Inn Air Crash and Fire, Wayne Township, Indiana (USFA-TR-014)*. Emmittsburg, Maryland (USA): US Fire Administration.
US Fire Administration (1991), *Technical Report Series: Major Ship Fire Extinguished by CO_2, Seattle, Washington (USFA-TR-058)*. Emmittsburg, Maryland (USA): US Fire Administration.
US Fire Administration (1994), *Technical Rescue Incident Report: The Derailment of the Sunset Limited, Big Bayot Canot, Alabama, September 22, 1993*. Emmitsburg, Maryland: US Fire Administration.
US Fire Administration (1996:16), *Technical Rescue Incident Report: The Crash of Two Subway Trains on the Williamsburg Bridge, June 5, 1995, New York City, New York*. Emmitsburg, Maryland: US Fire Administration.
US Fire Administration (1999), *Fire in the United States: 1987–1996*, 11th edition. Emmitsburg, Maryland: US Fire Administration.
US National Highway Traffic Safety Administration (NHTSA) (1999), *Traffic Safety Facts 1999: Occupant Protection (DOT HS 809 090)*. Washington, DC: National Center for Statistics & Analysis.
US National Transportation Safety Board (NTSB) (1985), *Safety Study – Airline Passenger Safety Education: A Review of Methods to Present Safety Information*. Report No. NTSB SS 85/09. Washington, DC: NTSB.
US National Transportation Safety Board (NTSB) (2001), *Safety Report: Survivability of Accidents Involving Part 121 US Air Carrier Operations, 1983 Through 2000 NTSB/SR-01/01*. Washington, DC: NTSB.

Vale, Gerald L., Anselmo, Joseph A. and Hoffman, Betty L. (1987), "Forensic Dentistry in the Cerritos Air Crash," *Journal of the American Dental Association*, volume 114 (May), pp. 661–664.

Valent, Paul (1995), "Survival Strategies: A Framework for Understanding Secondary Traumatic Stress and Coping in Helpers," C. R. Figley (ed.), *Compassion Fatigue: Coping with Secondary Traumatic Stress Disorder in Those Who Treat the Traumatized*, New York, Brunner/Mazel Publishers, pp. 21–50.

Vernon, Wesley (1994), "The Use of Chiropody/Podiatry Records in Forensic and Mass Disaster Identification," *Journal of Forensic Identification*, volume 44:1, pp. 26–40.

Vermylen, Y. (1991), "Some Legal Aspects of Mass Disasters," *Journal of Forensic Odonto-Stomatology*, volume 9:2, pp. 62–75.

Waeckerle, Joseph F. (1983), "The Skywalk Collapse: A Personal Response," *Annals of Emergency Medicine*, volume 12, p. 651 ff.

Walsh, Mike (1989), *Disasters, Current Planning and Recent Experience*. London: Edward Arnold Publishers.

Walsh, Mike (1990), "Taylor on Hillsborough: What Can We Learn?" in *Disasters*, volume 13:3, pp. 274–277.

Walton, M. and Lamb, J.P. (1980), *Raiders over Sheffield*. Sheffield: City Libraries.

Wang, J. (2000), "Analysis of Safety-Critical Software Elements in Offshore Safety Studies," *Disaster Prevention and Management*, volume 9:4, pp. 271–281.

Warnick, Allan J. (1987), "Dentists Aid in Identification of Crash Victims," *Journal of the Michigan Dental Association*, volume 69 (October), pp. 553–556.

Weedn, V.W. (1998), *The Unrealized Potential of DNA Testing*. US Dept of Justice, Office of Justice Programs, National Institute of Justice.

Welton, Bernard J.A.M. (1993), "Crisis Management Lecture: El Al Aircraft Carash/Bijlmer Disaster," *International Association of Airport & Seaport Police, 1993 Yearbook*, Port Coquitlam, British Columbia (Canada), pp. 11–16.

Wijkman, Anders and Timberlake, Lloyd (1984), *Natural Disasters: Acts of God or Acts of Man?* London: Earthscan Paperback.

Wilkinson, C. (1983), "Aftermath of a Disaster: The Collapse of the Hyatt Regency Hotel Skywalk," *American Journal of Psychiatry*, volume 140, pp. 1134–1139.

Williams, Henry (1964), in Grosser, George, Wechsler, Henry and Greenblatt, Milton (eds), *Threat of Impending Disaster*. Cambridge, Massachusetts: MIT Press.

Wilson, Harry C (2000), "Emergency Response Preparedness: Small group training. Part 1 – Training and Learning Styles," in *Disaster Prevention and Management*, volume 9:2, pp. 105–116.

Wood, Alan (2000), "Avoiding the Second Disaster," paper presented at Disaster Forum 2000, Edmonton, Canada, 2 November 2000

Wright J.E. (1977), "The Prevalence and Effectiveness of Centralized Medical Responses to Mass Casualty Disasters," in *Mass Emergencies*, volume 2, p.189 ff.

Zugibe, Frederick T., Constello, James, and Segelbacher, Joseph (1996), "The Horrors of Visual Misidentification," *Journal of Forensic Identification*, volume 46:4, pp. 403–406.

AUTHOR INDEX

SUBJECT INDEX